DESIGN AUTOMATION FOR TIMING-DRIVEN LAYOUT SYNTHESIS

THE KLUWER INTERNATIONAL SERIES IN ENGINEERING AND COMPUTER SCIENCE

VLSI, COMPUTER ARCHITECTURE AND DIGITAL SIGNAL PROCESSING
Consulting Editor
Jonathan Allen

Latest Titles

Monte Carlo Device Simulation: Full Band and Beyond, Karl Hess, editor
   ISBN: 0-7923-9172-1
The Design of Communicating Systems: A System Engineering Approach,
   C. J. Koomen
   ISBN: 0-7923-9203-5
Parallel Algorithms and Architectures for DSP Applications,
   M. A. Bayoumi, editor
   ISBN: 0-7923-9209-4
Digital Speech Processing: Speech Coding, Synthesis and Recognition
   A. Nejat Ince, editor
   ISBN: 0-7923-9220-5
Sequential Logic Synthesis, P. Ashar, S. Devadas, A. R. Newton
   ISBN: 0-7923-9187-X
Sequential Logic Testing and Verification, A. Ghosh, S. Devadas, A. R. Newton
   ISBN: 0-7923-9188-8
Introduction to the Design of Transconductor-Capacitor Filters,
   J. E. Kardontchik
   ISBN: 0-7923-9195-0
The Synthesis Approach to Digital System Design, P. Michel, U. Lauther, P. Duzy
   ISBN: 0-7923-9199-3
Fault Covering Problems in Reconfigurable VLSI Systems, R.Libeskind-Hadas,
   N. Hassan, J. Cong, P. McKinley, C. L. Liu
   ISBN: 0-7923-9231-0
High Level Synthesis of ASICs Under Timing and Synchronization Constraints
   D.C. Ku, G. De Micheli
   ISBN: 0-7923-9244-2
The SECD Microprocessor, A Verification Case Study, B.T. Graham
   ISBN: 0-7923-9245-0
Field-Programmable Gate Arrays, S.D. Brown, R. J. Francis, J. Rose,
   Z.G. Vranesic
   ISBN: 0-7923-9248-5
Anatomy of A Silicon Compiler, R.W. Brodersen
   ISBN: 0-7923-9249-3
Electronic CAD Frameworks, T.J. Barnes, D. Harrison, A.R. Newton,
   R.L. Spickelmier
   ISBN: 0-7923-9252-3
VHDL for Simulation, Synthesis and Formal Proofs of Hardware, J. Mermet
   ISBN: 0-7923-9253-1
Wavelet Theory and its Applications, R. K. Young
   ISBN: 0-7923-9271-X
Digital BiCMOS Integrated Circuit Design
   S.H.K. Embabi, A. Bellaouar, M.I Elmasry
   ISBN: 0-7923-9276-0

DESIGN AUTOMATION FOR TIMING-DRIVEN LAYOUT SYNTHESIS

by

Sachin S. Sapatnekar

Sung-Mo (Steve) Kang

University of Illinois, Urbana-Champaign

KLUWER ACADEMIC PUBLISHERS
Boston/Dordrecht/London

Distributors for North America:
Kluwer Academic Publishers
101 Philip Drive
Assinippi Park
Norwell, Massachusetts 02061 USA

Distributors for all other countries:
Kluwer Academic Publishers Group
Distribution Centre
Post Office Box 322
3300 AH Dordrecht, THE NETHERLANDS

Library of Congress Cataloging-in-Publication Data

Sapatnekar, Sachin S., 1967-
    Design automation for timing-driven layout synthesis / by Sachin
  S. Sapatnekar, Sung-Mo (Steve) Kang.
      p. cm. -- (The Kluwer international series in engineering and
  computer science ; 0198)
    Includes bibliographical references and index.
    ISBN 0-7923-9281-7
    1. Metal oxide semiconductors, Complementary--Design and
  construction--Data processing.   2. Integrated circuits--Very large
  scale integration--Design and construction--Data processing.
  3. Computer-aided design. I. Kang, Sung-Mo, 1945- II. Series:
  Kluwer international series in engineering and computer science ;
  SECS 0198.
  TK7871 . 99 . M44S37   1992
  621 . 39 ' 5--dc20   92-29798
                                                                                                   CIP

Copyright © 1993 by Kluwer Academic Publishers

All rights reserved. No part of this publication may be reproduced, stored in a retrieval system or transmitted in any form or by any means, mechanical, photo-copying, recording, or otherwise, without the prior written permission of the publisher, Kluwer Academic Publishers, 101 Philip Drive, Assinippi Park, Norwell, Massachusetts 02061.

Printed on acid-free paper.

Printed in the United States of America

"'Tis the good reader that makes the good book."

- Ralph Waldo Emerson

Contents

List of Figures — xiii

Acknowledgements — xix

1 Introduction — 1

   1.1 The Process of IC Design 2

   1.2 Layout Styles 4

   1.3 Timing-driven Layout 9

   1.4 Outline of the Book 11

2 Delay Estimation — 13

   2.1 Introduction 13

   2.2 Micromodeling - The RC Model 16

      2.2.1 Introduction 16

	2.2.2	Definitions : RC Trees and RC Meshes	16
	2.2.3	RC Representation of a CMOS circuit	18
	2.2.4	The Elmore Time Constant	21
	2.2.5	Penfield-Rubenstein Bounds	28
2.3	Macromodeling		31
	2.3.1	The Table Look-up Method	33
	2.3.2	The Analytical Method	40
2.4	Worst-case Delay Estimation		50
2.5	Delay Calculation at the Circuit Level		52
	2.5.1	Combinational Subnetwork Extraction	52
	2.5.2	The PERT Method	56
2.6	Posynomial Delay Modeling		60
2.7	A Case Study: iCONTRAST's Timing Analyzer		62
	2.7.1	Transistor-level Micromodeling	63
	2.7.2	The Delay Estimation Algorithm	66
	2.7.3	A Comparison with SPICE	76
2.8	Summary		77

3 Transistor Sizing Algorithms : Existing Approaches 81

3.1 Introduction 81

3.2	The TILOS Algorithm		85
	3.2.1	The Area Model	85
	3.2.2	The TILOS Delay Model	86
	3.2.3	The TILOS Optimizer	89
	3.2.4	Sensitivity Computation	91
3.3	The Method of Feasible Directions (MFD) Algorithm		93
	3.3.1	Description of the Algorithm	93
	3.3.2	Practical Implementational Aspects	96
3.4	Lagrangian Multiplier Approaches		100
	3.4.1	Early Approaches	101
	3.4.2	Marple's Approach	102
3.5	Two-step Optimization		106
3.6	Other Approaches		108
3.7	Summary of Previous Approaches		109

4 A Convex Programming Approach to Transistor Sizing   113

4.1	Introduction		113
4.2	The Convex Programming Algorithm		116
	4.2.1	Finding the Center of the Polytope	122
	4.2.2	Rank-one Updates	125

		4.2.3	One-dimensional Line Search 130
		4.2.4	Generation of Hyperplanes 131
		4.2.5	Termination Criterion 132
	4.3	Experimental Results 133	
	4.4	Summary 138	

5 Global Routing Using Zero-one Integer Linear Programming — 141

- 5.1 Introduction 141
- 5.2 Extracting Global Routing Information 146
- 5.3 Global Routing Phases 148
  - 5.3.1 Phase One Routing 150
  - 5.3.2 Phase Two Routing 154
  - 5.3.3 Phase Three Routing 160
- 5.4 Global Routing on Medium-sized Arrays 173
- 5.5 Application to Custom Logic Layout 179
- 5.6 Handling Very Large Circuits 182
- 5.7 Runtime Complexity 185
- 5.8 Conclusion 188

6 Timing-driven CMOS Layout Synthesis — 191

- 6.1 Introduction … 191
- 6.2 A Methodology for Designing CMOS Standard Cells — 193
  - 6.2.1 Introduction … 193
  - 6.2.2 Design Objective … 195
  - 6.2.3 Basic Assumptions … 197
  - 6.2.4 Propagation Delays in CMOS Gates … 197
  - 6.2.5 Consideration of Noise Margins … 204
  - 6.2.6 Worst-case Simulations and Standard Cell Design … 206
  - 6.2.7 Conclusion … 211
- 6.3 The Metal-Metal Matrix (M^3) Layout Style for Two-level Technologies … 213
- 6.4 iCGEN : A CMOS Layout Synthesis System for Three-level Metal Technology … 216
  - 6.4.1 Outline of iCGEN … 216
  - 6.4.2 Layout Platform … 222
  - 6.4.3 Mapping from Triple Metal Layers to Double Metal Layers … 227
  - 6.4.4 Determination of Optimal Row Height … 227

6.4.5	Channelless Logic Cell Placement	230
6.4.6	Routing of Intercell Signals	231
6.4.7	Logic Cell Layout Generation	234
6.4.8	Circuit Tuning Using Convex Optimization	236
6.4.9	Experimental Results	238
6.4.10	Summary	244

Bibliography   **247**

Index   **267**

List of Figures

1.1 A typical IC design process. 3

1.2 Organization of the book. 11

2.1 An example of a composite cell. 14

2.2 Illustration of the transition delay. 15

2.3 An RC tree. 17

2.4 An RC mesh. 18

2.5 An RC transistor model. 19

2.6 RC Models for interconnect. 20

2.7 Cross-coupling capacitances between interconnect lines. 21

2.8 Elmore's definition of delay. 22

2.9 Replacing capacitors by current sources. 25

2.10 The Penfield-Rubenstein bounds. 29

2.11 Identification of channel-connected components. 32

2.12 The five basic primitives for Rao et al.'s approach. . . 35

2.13 (a) Inverter circuit (b) Input signal (c) Output waveform. 42

2.14 Flowchart for inverter delay time calculation. 48

2.15 (a) Two-input NAND gate (b) The equivalent inverter. 49

2.16 An example with two paths between output and ground. 51

2.17 Set-up time requirements. 56

2.18 The PERT technique. 57

2.19 Circuit delays obtained from the PERT technique. . . 59

2.20 The physical parameters associated with a transistor. . 64

2.21 (a) An inverter undergoing a fall transition (b) Its equivalent RC network. 65

2.22 Graphical illustration of t_f and Δ_f. 67

2.23 Extending the LRP to a tree. 72

2.24 (a) iCONTRAST's timing analyzer vs. SPICE (b) An Elmore delay based timing analyzer without iCONTRAST's enhancements vs. SPICE 79

3.1 (a) A chain of three inverters (b) Effect of transistor sizes on delay for the three-inverter chain. 83

3.2 (a) A sample pulldown network (b) Its RC representation. 88

LIST OF FIGURES

3.3 Sensitivity calculation in TILOS. 91

3.4 Nondifferentiability of the *max* function. 96

3.5 The convex hull of five points. 97

3.6 Computing the generalized gradient. 99

4.1 Example illustrating the convex programming algorithm. 121

4.2 (a) iCONTRAST's timing analyzer vs. SPICE (b) An Elmore delay based timing analyzer without iCONTRAST's enhancements vs. SPICE 140

5.1 An overview of the SOG global router. 146

5.2 Possible vertical path segments for inter-row routing. . 151

5.3 Possible straight-line metal1 paths. 156

5.4 Possible straight-line metal2 paths. 157

5.5 An example of a loop avoidance constraint. 159

5.6 A possible route to be formed in Phase Three routing. 161

5.7 Steps in solving a 2×16 problem. 164

5.8 A potential via-GRC. 165

5.9 Representation of the 2×2 macro-GRC routing problem. 165

5.10 Possible routes in nets with one macro-node. 167

5.11 Possible routes in nets with two macro-nodes. 169

5.12 Possible routes in nets with three macro-nodes. 170

5.13 Possible routes in nets with four macro-nodes. 171

5.14 Boundary and via capacities of 2 × 2 macro-GRCs. . . 172

5.15 Routing area showing the routing result of circuit **i10**. 175

5.16 The routing process for net G of circuit **i10**. 175

5.17 Percentage of nets routed vs. available horizontal and vertical tracks. 177

5.18 Packing densities of **c162** and reported CMOS chips. . 179

5.19 Packing densities of **c1318** and reported BiCMOS chips.181

5.20 Placement and global routing of a four-bit adder. . . . 183

5.21 Runtime complexity of the global router. 187

6.1 A typical layout of CMOS standard cell NAND2. . . . 195

6.2 50% point propagation delays, τ_n and τ_p, in a CMOS inverter with a fanout of 3. 198

6.3 Equivalent circuits for (a) NAND3 and (b) NOR3 with two inputs set at V_{DD} and 0, respectively. 200

6.4 (a) CMOS inverter circuit. (b) Its input-output signal pair. 201

6.5 (a) The propagation delay (ns) of an average logic gate. (b) The worst-case noise margin/V_{DD} vs. the aspect ratio $R \triangleq W_p/W_n$. 207

LIST OF FIGURES

6.6 dc transfer characteristics of INR with $W_p/W_n = 26/26$. (a) High-current n-channel and low-current p-channel parameters. (b) Low-current n-channel and high-current p-channel parameters. 208

6.7 dc transfer characteristics of INR with $W_p/W_n = 35/17$. (a) High-current n-channel and low-current p-channel parameters. (b) Low-current n-channel and high-current p-channel parameters. 209

6.8 (a) The average-gate propagation delay (ns). (b) The delay time-chip area product (10^{-12}s.m^2) vs. the total channel width $W = W_n + W_p$. 210

6.9 The worst-case propagation delay, τ, of an "average" standard cell gate versus capacitive loading C_L. ... 212

6.10 Two layout structures for the inverter circuit. 214

6.11 Layout example using M^3. 215

6.12 An overview of the layout system. 217

6.13 Base cell of the logic platform. 219

6.14 Cell placement without channels. 231

6.15 Automatic layout of industrial circuit 1. 241

6.16 Automatic layout of industrial circuit 2. 242

6.17 Area vs. speed for a complex two-bit adder circuit. ... 243

Acknowledgements

Many colleagues have provided us with support, advice and commentary on this book. In particular, the authors would like to thank Dr. V. B. Rao of IBM Corporation and Prof. P. M. Vaidya of the University of Illinois at Urbana-Champaign for their contributions to the convex programming based transistor sizing algorithm. We acknowledge the contributions of Dr. Hau-Yung Chen of Texas Instruments to the macromodeling work and iCOACH, Mr. Ngee Lek of the Singapore Economic Development Board to the global router, and Mr. Richard Thaik of National Semiconductor Corporation for his efforts in putting iCGEN together. We are also indebted to Dr. J. P. Fishburn and Dr. A. E. Dunlop of AT&T Bell Laboratories for their helpful comments and assistance with regard to our work in transistor sizing. Finally, we would like to thank the members of the Digital and Analog Circuits Group at the Coordinated Science Laboratory, University of Illinois at Urbana-Champaign, and Prof. T. N. Trick, the Head of the Department of Electrical and Computer

Engineering at the University of Illinois at Urbana-Champaign.

We would like to express our appreciation to Lilian Beck for proofreading the manuscript, and Bob MacFarlane for designing the cover.

Special thanks are also due to the family of the first author, namely, Suresh, Sudha and Suneeti Sapatnekar, and to the wife of the second author, Myoung-A. (Mia) Kang.

Much of the work described herein was supported by the Joint Services Electronics Program, Semiconductor Research Corporation, the State of Illinois Technology Challenge Grant, and the Center for Advanced Study at the University of Illinois at Urbana-Champaign.

DESIGN AUTOMATION FOR TIMING-DRIVEN LAYOUT SYNTHESIS

Chapter 1

Introduction

Moore's law [Noy77], which predicted that the number of devices integrated on a chip would be doubled every two years, was accurate for a number of years. Only recently has the level of integration begun to slow down somewhat due to the physical limits of integration technology. Advances in silicon technology have allowed IC designers to integrate more than a few million transistors on a chip; even a whole system of moderate complexity can now be implemented on a single chip.

To keep pace with the increasing complexity in very large scale integrated (VLSI) circuits, the productivity of chip designers would have to increase at the same rate as the level of integration. Without such an increase in productivity, the design of complex systems might not be achievable within a reasonable time-frame.

The rapidly increasing complexity of VLSI circuits has made design automation an absolute necessity, since the required increase in productivity can only be accomplished with the use of sophisticated design tools. Such tools also enable designers to perform trade-off analyses of different logic implementations and to make well-informed design decisions.

1.1 The Process of IC Design

A typical IC design process, shown in Figure 1.1, is composed of the four following categories:

- *System (behavioral) design* is the process of defining the circuit functionality and the input-output behavior. This level of specification may be expressed in terms of a flowchart or in terms of a high-level hardware description language (HDL).
- *Logic design* is the process of transforming a high-level description of a complex function into a netlist of technology-independent logic elements such as NAND gates, NOR gates, inverters, AOI gates and latches. This process helps to ensure that minimal logic is used to implement the function that was earlier defined in a high-level language.
- *Circuit design* transforms the basic logic components into networks of transistors and interconnects.

1.1. THE PROCESS OF IC DESIGN

- *Layout design* creates geometrical shapes on various mask layers, which correspond to a silicon implementation of the circuit.

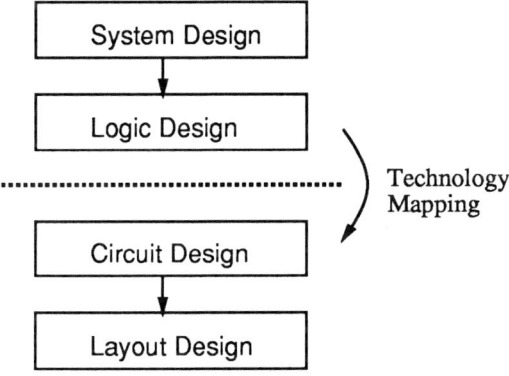

Figure 1.1: A typical IC design process.

Although these steps are interrelated, each has its primary goals. At the system design level, the goal is to provide a complete and precise functional description. Logic-level design attempts to implement the behavioral descriptions using logic gates or other entities. At the circuit level, the aim is to optimize the timing and to reduce the power consumption, and the objective of layout design is to realize circuit functions with a high packing density.

System design and logic design are often combined into logic synthesis. The logic synthesis process translates a high-level functional description into a gate-level representation in three steps:

- global minimization of Boolean equations
- factorization of Boolean equations

- local transformation of logic structures.

The output of logic synthesis is a description of the circuit in terms of logic blocks, such as NAND gates, NOR gates, and flip-flops. The gate structure provided by logic synthesis is considered to be technology-independent.

This technology-independent gate structure is mapped to a specific technology during the process of circuit design and layout. The transition from a technology-independent process to a technology-dependent process, i.e., from logic synthesis to the circuit and layout design phases, is called *technology mapping*. This procedure is carried out using a specified layout style, such as PLA, gate array, sea-of-gates, standard cell or full-custom design implementations. These styles are discussed in the next section.

1.2 Layout Styles

Programmable Logic Array (PLA) Approach

Programmable logic arrays (PLAs) provide a flexible and efficient way of synthesizing arbitrary combinational functions in a regular structure. Due to the easy access of PLA CAD tools that provide a fast turnaround time, the PLA has been used frequently in circuit design. Most PLA generators incorporate optimization techniques

1.2. LAYOUT STYLES

to perform logic minimization, partitioning, and row/column foldings achieve compact AND/OR planes. However, PLA layouts can be area-efficient only when the number of product terms is small and the logic is properly partitioned; otherwise, the size of the PLA tends to grow quadratically with respect to the number of input terms. Another limitation is that large parasitic capacitances tend to limit the speed of the PLA, although some approaches alleviate such problems by the judicious use of buffers. Very large PLAs are, therefore, impractical in view of area and circuit performances.

Gate Array Approach

Gate array and standard cell approaches are often referred to as *semi-custom* design styles, since they use predefined patterns or cells as opposed to the full-custom approach which employs transistors as the basic elements. A gate array consists of a chip in which a predefined pattern of transistors is arranged in the form of a matrix. These transistors have already undergone all of the stages of the chip fabrication process except for the final metallization steps. Since every chip contains the same predefined layout of transistors independent of its final interconnection, the turn-around time from the design stage to chip manufacturing is short. The part of the manufacturing process that is unique to each design is the metallization stage in which a particular design pattern (the personality) is deposited on the chip. Since this is only a small fraction of the total process time, the time required to translate a design into a product is significantly reduced.

There are, however, some drawbacks. Gate arrays shorten and simplify the design process at the expense of area efficiency due to the use of a universal structure that must accommodate different applications. As a result, many components in the structure may not be used when actual circuits are built, and a considerable amount of area is wasted. Another problem with the gate array approach is that the transistor patterns are predefined and therefore the transistors cannot be tuned to the specific application. Hence its performance is inferior to that of fully customized chips.

Sea-of-gates Approach

The sea-of-gates array, also known as the continuous gate array, channelless array, or gate forest, is an enhancement of the gate array structure described above. The sea-of-gates is *channelless*, i.e., there are no predefined channels for routing; routing is carried out over core cells instead. Moreover, the gate isolation concept, as opposed to oxide isolation, is used, which means that the basic cell layout includes no gap in the diffusion layer; electrical isolation is performed

1.2. LAYOUT STYLES 7

by connecting a transistor gate to the appropriate supply (V_{DD} or V_{SS}) rail.

Standard Cell Approach

Standard cell approach is currently the mainstay in Application Specific Integrated Circuits (ASICs). This approach involves the use of a library of basic functional elements, each of which has been fully characterized.

The cell library approach has the advantage of greatly simplifying the automated synthesis process because it separates the synthesis system from the details of cell layout issues. The cell library presents models for individual cells, which are useful for performing circuit and timing analyses.

A crucial issue in the cell library approach is the size of the library. If the library is too small, much time is spent in converting the logic into a format that can be supported by the small library. On the other hand, if the library size is too large, the issues of database maintenance, pattern matching and searching become significant [Kah87]. Moreover, the useful life of a library is relatively short as dictated by the lifetime of the technology in use [KKL87]. For these reasons, the cell libraries tend to remain relatively small in size.

The prevalent use of complex gates such as AOI or OAI further complicates the library issue. As shown in Table 1.1 [DGR87], as

Table 1.1: Number of (s,p)-gates.

(s,p)	1	2	3	4	5	6
1	1	2	3	4	5	6
2	2	7	18	42	90	186
3	3	18	87	396	1677	6877
4	4	42	396	3503	28435	222943
5	5	90	1677	28435	425803	6084393
6	6	186	6877	222943	6084393	154793519

many as 3,503 different complex gates can be configured for $(s,p) = (4,4)$, where the gates are constrained to have at most s transistors from output to ground and p transistors from output to power supply. This number dramatically increases to 425,803 for $(s,p) = (5,5)$ and 154,793,519 for $(s,p) = (6,6)$. It is apparent that a moderately sized library cannot support all of the possible circuit configurations for complex gates.

The other problem with standard cell approach is that even if individual cells in the library are nearly optimal in performance and in terms of compactness of the layout, the whole circuit is often suboptimal after all cells are put together. For flexibility, many standard cell libraries contain multiple versions of some cells with different driving powers.

1.3 Timing-driven Layout

To ensure that high-quality designs are produced, a CAD system must take two important issues into consideration while designing a circuit:

- Layout efficiency : producing a compact circuit layout.
- Performance : satisfying the timing specifications dictated by the clocking scheme.

With the increasing drive for high-performance chips, timing-driven layout has become the watchword.

For a given process technology, several issues need to be addressed in regard to achieving this objective at the circuit design stage:

Logic Design

In many cases, a reduction in the number of stages (gates) between an input and an output node can reduce the circuit area and delay. Such an optimization is usually made during the logic synthesis stage. This reduction is not, however, guaranteed to reduce the circuit delay. For example, when a large load needs to be driven, as in the case of buffer design, it has been illustrated in [MC80] that the introduction of a certain number of additional stages could reduce the overall circuit delay. In this book, we assume that a technology-independent logic-level circuit description is provided to us, and the objective is to perform the technology mapping in an optimal way. The logic synthesis stage is usually performed before the technology mapping stage. Hence, we do not address this issue in this book.

Placement

As device geometries continue to shrink, interconnect delays become increasingly significant. As a result, the reduction of interconnect wire length, which heavily influences the interconnect delay, has become increasingly important. Therefore, it becomes important to place the building blocks or modules in a design in such a way that the wire lengths are minimal. Various algorithms have been developed to solve the timing-driven placement problem [SCK91, GVL91].

Transistor Size Optimization

For a given circuit topology, the delay of a circuit can be controlled by varying the sizes of transistors in the circuit. Roughly speaking, the circuit delay can usually be reduced by increasing the sizes of certain transistors in the circuit, though the precise relation is more complex, as illustrated in Chapter 3. However, this involves an area overhead, and a careful tradeoff must be made between the area and the delay.

1.4 Outline of the Book

In this book, we address the problem of performance-driven layout for a given circuit topology, with particular reference to the algorithms used in layout synthesis systems. The overall flow of the book is depicted in Figure 1.2.

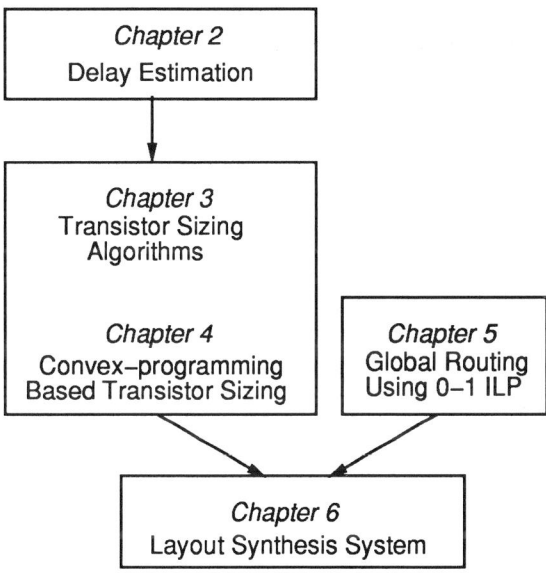

Figure 1.2: Organization of the book.

Chapter 2 provides an overview of fast algorithms for worst-case delay estimation, along with a case study of a practical algorithm that is used for transistor sizing. Next, a survey of various techniques for transistor sizing is provided in Chapter 3. Chapter 4 describes a convex programming-based approach to solving the transistor sizing problem which is implemented in the program iCONTRAST.

In Chapter 5, a global routing algorithm using 0-1 Integer Linear Programming (ILP) techniques is presented. Finally, in Chapter 6, we present two approaches to performance-driven layout. The first of these is an approach to designing standard cells for a standard cell library. The second describes an automated timing-driven layout synthesis system, iCGEN, that uses a suite of program modules to obtain a compact layout that meets user-specified timing requirements.

Chapter 2

Delay Estimation

2.1 Introduction

Finding the delay of a digital circuit accurately is an important part of the design and verification process. Various levels of simulation can be used, depending on the accuracy desired, and the amount of CPU time that is affordable.

Of all techniques currently used to simulate a circuit, circuit-level simulation [CL75] provides the highest degree of accuracy. This method attempts to solve the circuit equations, i.e., the KCL and KVL equations in conjunction with the device characteristics, to find the voltage at every node and the current through every branch of the circuit. The process involves solving systems of nonlinear differential equations. The chief drawbacks of this method are the high computational cost and extensive memory requirements. In addition, due to various problems associated with nonlinear algebraic and differential equations, the convergence of the solution process is not always guaranteed.

To circumvent these disadvantages, techniques such as timing simulation, logic simulation and mixed-mode simulation are used to compute waveforms of interest using various approximations, sacrificing accuracy for fast computation.

Often, a rough estimate of the circuit delay can suffice for solving macroscopic design issues, and one need not pay attention to precise voltage or current waveforms. The following alternatives can be used:

- The *micromodeling* approach replaces each transistor by an equivalent model. Our discussion in this book is based on the most common form of transistor modeling, namely RC (resistor-capacitor) modeling.

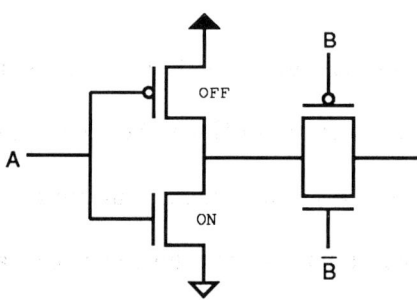

Figure 2.1: An example of a composite cell.

- The *macromodeling* approach uses a higher level of abstraction; the fundamental unit that is modeled could be a gate or a

2.1. INTRODUCTION

functional block. This book deals with macromodels at the gate level. This level contains basic gates such as inverters, NANDs, NORs, XORs and complex gates such as AOI and OAI gates, or certain composite cells, such as that in Figure 2.1. Such a composite cell is used as a primitive in timing analysis [ROT89].

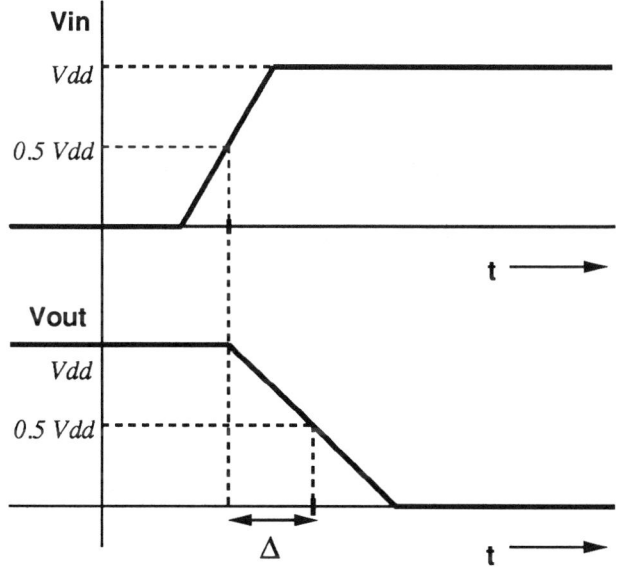

Figure 2.2: Illustration of the transition delay.

Before proceeding further, it would be helpful to define the delay associated with a transition. Consider the input and output waveforms shown in Figure 2.2. We define the transition delay as the amount of time required by the output waveform to cross the 50% threshold, after the input waveform has crossed its 50% threshold.

In Figure 2.2, Δ represents the delay associated with the transition.

2.2 Micromodeling - The RC Model

2.2.1 Introduction

The use of resistors and capacitors for transistor level modeling has several advantages over more detailed simulation procedures, although it involves a loss in accuracy. In addition to the major reduction in the computational complexity, the RC representation often offers a simple closed-form formula for the delay of a circuit. Such simple formulæ are particularly useful for delay optimization, since it is far easier, in terms of tractability and computational cost, to perform mathematical operations on this relatively simple delay expression. Another benefit provided by this technique is that the effects of interconnect delay can easily be incorporated by augmenting the RC network that describes the MOS transistor netlist, with additional resistors and capacitors that represent interconnect lines.

2.2.2 Definitions : RC Trees and RC Meshes

An *RC tree* may be defined recursively as follows [RPH83]. There are three primitive elements:

- A capacitor between ground and another node is an RC tree.

2.2. MICROMODELING - THE RC MODEL

- A resistor between two nonground nodes is an RC tree.
- An RC line, in the configuration with no dc path to ground, is an RC tree.

Finally, any two RC trees with a common ground node, and one nonground node from each tree connected together, form a new RC tree.

The following properties of an RC tree follow from this definition:

(1) Resistor loops are not permitted; therefore, all of the resistors form a topological tree that does not include the ground node.
(2) All of the capacitors in an RC tree are connected to ground.
(3) One of the nonground nodes of the tree is assumed to be the input, and one or more nodes the outputs. As a consequence of (1) and (2), there is a unique path through the resistive path of the network from any nonground node to the input node.

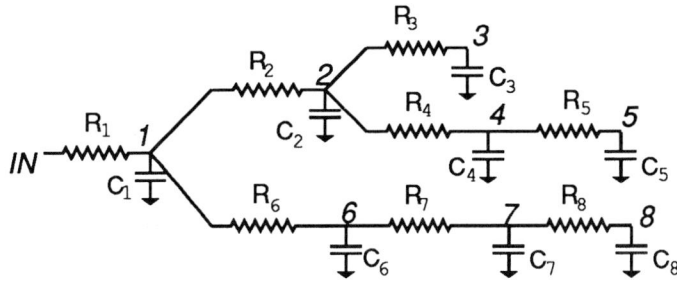

Figure 2.3: An RC tree.

An example of an RC tree is shown in Figure 2.3. The input node of the network is the one marked *"IN"*. Note that this tree satisfies the three properties listed above.

An *RC mesh* has the same three primitive elements as an RC tree. The difference lies in the recursive definition: any two RC meshes with a common ground node and *one or more* nonground nodes from each tree connected together form a new RC mesh.

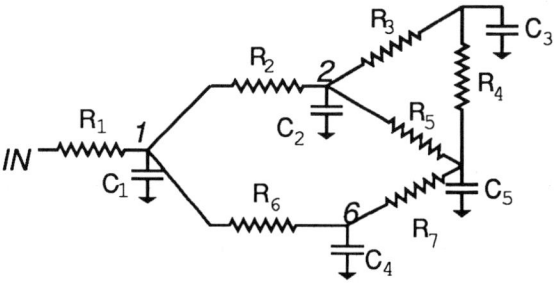

Figure 2.4: An RC mesh.

In essence, the difference between an RC tree and an RC mesh is that resistor loops are permitted in RC meshes. An example of an RC mesh is shown in Figure 2.4.

2.2.3 RC Representation of a CMOS circuit

In this section, we examine how a CMOS circuit may be represented by a set of RC networks. The techniques discussed later in this chapter may be used to find an approximation to the delay of each

2.2. MICROMODELING - THE RC MODEL

RC network. These methods can be used in conjunction with the procedure in Section 2.5 to find an approximation to the circuit delay.

The basic elements of a CMOS integrated circuit are transistors (n-channel and p-channel), interconnect lines, and resistors and capacitors. In order to see how such a circuit could be modeled by an RC network, we examine how transistors and interconnect lines may be represented by RC models.

RC Model of a Transistor

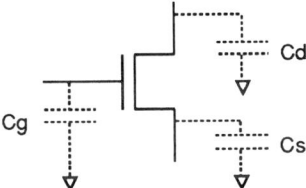

Figure 2.5: An RC transistor model.

A MOS transistor is modeled as a voltage-controlled switch with on-resistance R_{on} between drain and source and three grounded capacitances C_d, C_s, and C_g from the drain, source, and gate terminals, respectively, to ground. This is illustrated in Figure 2.5.

RC Models for Interconnect Lines

Under current operating frequencies, inductive effects in interconnect lines are considered to be negligible. Hence RC models, rather than

RLC models, for interconnect are considered adequate.

Strictly speaking, interconnect is a distributed system that is better represented by distributed resistors and capacitors, rather than by lumped elements. However, since lumped elements are easier to analyze, such models are favored over distributed models, especially for circuits where the operating frequency is not very high. The most obvious way to approximate a uniform distributed interconnect segment by lumped elements is by using a cascade of sections, as shown in Figure 2.6(a).

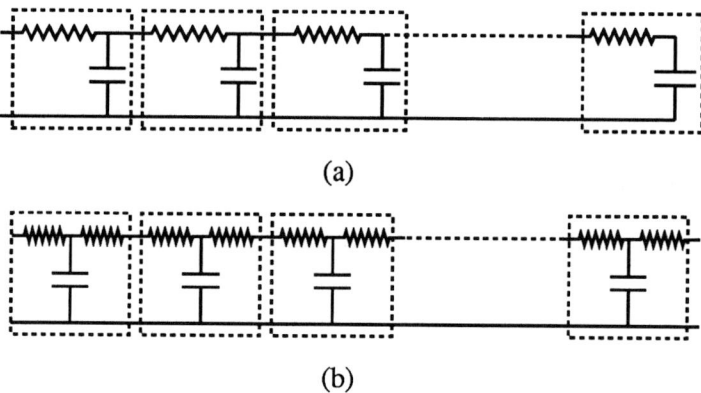

Figure 2.6: RC Models for interconnect.

The cascaded symmetric "T-sections", illustrated in Figure 2.6(b), produce a better representation of interconnect behavior [Sak83]. As the number of segments is increased, the step response of this approximation converges faster (as compared to the approximation in Figure 2.6(a)) thatn the response of a distributed line.

2.2. MICROMODELING - THE RC MODEL

To account for the effects of cross-coupling capacitances between lines, which may cause crosstalk noise, each interconnect line is considered to be a dc source. Since capacitive coupling to an ideal dc source is equivalent to a capacitive coupling to ground, these coupling capacitances may be modeled by capacitances to ground, as shown in Figure 2.7. The rationale for this approximation is that when all capacitances in the RC network are connected to the ground node, we have an RC mesh, whose delay can easily be found using the methods described in the following sections.

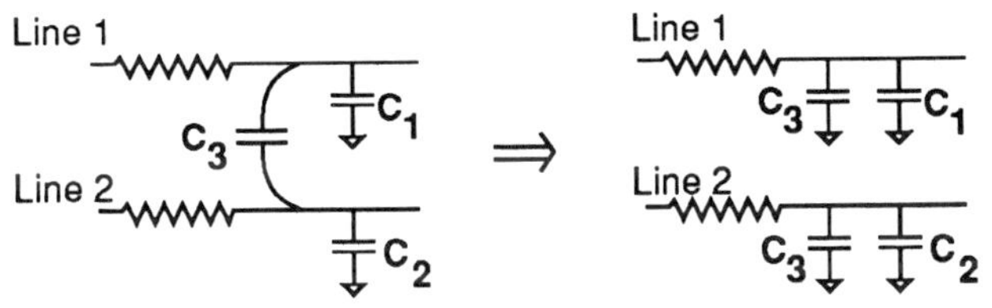

Figure 2.7: Cross-coupling capacitances between interconnect lines.

A good overview of interconnect modeling is provided in [Wya87].

2.2.4 The Elmore Time Constant

Elmore's Definition of Delay

Figure 2.8(a) illustrates the output waveform, $e(t)$, of an RC network, in response to a unit step excitation at its input. The delay time should be measured from $t = 0$, when the input step transition occurs, to the time at which the transient voltage reaches the 50%

point. Elmore suggested [Elm48] that the center-of-area of the region under the curve $e'(t)$, shown in Figure 2.8(b), would serve as a reasonable estimate to the delay, that is,

$$T_D = \int_0^\infty te'(t)dt. \qquad (2.1)$$

The equation for the centroid takes this simple form since

$$\int_0^\infty e'(t)dt = e(\infty) - e(0) = 1. \qquad (2.2)$$

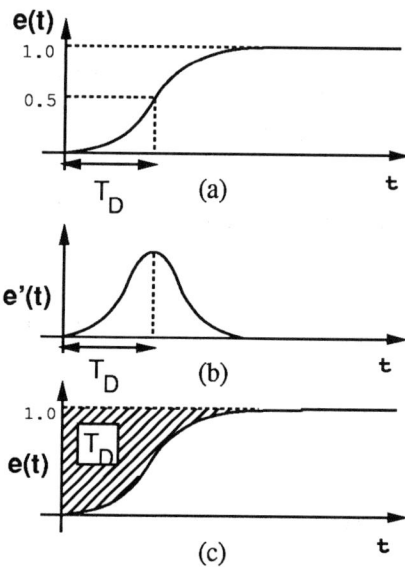

Figure 2.8: Elmore's definition of delay.

Since $e(t)$ is the step response of the rested system, the impulse response of the system is given by $e'(t)$. Hence, Equation (2.1) also

2.2. MICROMODELING - THE RC MODEL

has the interpretation of being the first moment, or mean, of the impulse response.

Moreover, the quantity T_D is also the area above the step response, as shown in Figure 2.8(c). This can be seen by performing integration by parts, as shown below:

$$\begin{aligned} T_D &= \int_0^\infty t e'(t) dt \\ &= \int_0^\infty [1 - e(t)] \, dt - t[1 - e(t)] \Big|_0^\infty \\ &= \int_0^\infty [1 - e(t)] \, dt \end{aligned} \quad (2.3)$$

Here, we make use of the fact that $\lim_{t \to \infty} t[1 - e(t)] = 0$ since $e(t) \to 1$ exponentially as $t \to \infty$.

The Elmore delay is a good approximation for the dominant time constant of the step response [Wya87]. This approximation is adequate for state-of-the-art CMOS circuits, where an RC tree model is sufficient for circuit modeling since RC trees have a single dominant time constant, which is estimated by the Elmore time constant. However, for bipolar circuits, and for MOS circuits at high speeds at which inductive effects become noticeable, there may be more than one dominating time constant; hence more sophisticated techniques such as asymptotic waveform evaluation (AWE) [PR90] and an extension of Elmore's delay presented in [GZ92] may be used.

Elmore Delay Through RC Networks

We consider the case of an RC mesh, i.e., resistor networks where there is a capacitor between every node and ground. We consider the case in which all resistors and capacitors are lumped elements.

For a given RC network with n grounded capacitors, the node conductance matrix [CL75] has the following form

$$\mathbf{G} = \begin{bmatrix} G_{1,1} & G_{1,2} & \cdots & -G_{1,n} \\ -G_{2,1} & G_{2,2} & \cdots & -G_{2,n} \\ \vdots & \vdots & \ddots & \vdots \\ -G_{n,1} & G_{n,2} & \cdots & G_{n,n} \end{bmatrix} \qquad (2.4)$$

where

$G_{i,j}$: branch conductance between nodes i and j ($G_{i,j} = G_{j,i}$)
$G_{i,i}$: sum of all branch conductances connected to node i.

We also define the capacitance vector

$$\mathbf{C} = \begin{bmatrix} C_1 \\ C_2 \\ \vdots \\ C_n \end{bmatrix} \qquad (2.5)$$

where C_i is the capacitance at node i. Here, the coupling capacitance effects are assumed to be neglected, or lumped into individual self-capacitances.

2.2. MICROMODELING - THE RC MODEL

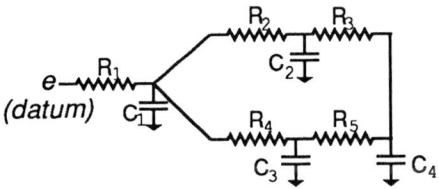

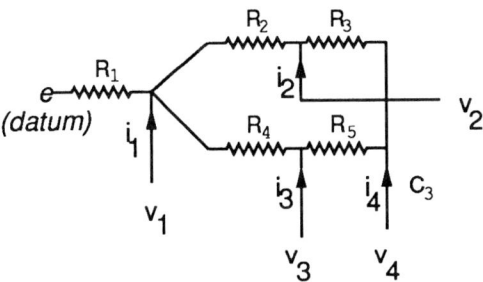

Figure 2.9: Replacing capacitors by current sources.

Theorem [Wya87] : The Elmore delay, T_{Di}, to the ith node of an RC network is given by

$$\mathbf{T_D} = \mathbf{G}^{-1}\mathbf{C} \qquad (2.6)$$

where $\mathbf{T_D} \in \mathbf{R}^n$ and $T_{Di} = (\mathbf{T_D})_i$.

Proof : Replace all capacitors to ground by equivalent controlled current sources, as shown by the example in Figure 2.9. The current flowing into node i due to such a source is

$$-C_i \frac{dV_i}{dt} \qquad (2.7)$$

where V_i is the voltage at node i.

The network now consists entirely of resistors and current sources.

Taking the input node as the datum, its nodal equations [CL75] for t > 0 may be written as

$$\begin{bmatrix} G_{1,1} & \cdots & -G_{1,n} \\ -G_{2,1} & \cdots & -G_{2,n} \\ \vdots & \ddots & \vdots \\ -G_{n,1} & \cdots & G_{n,n} \end{bmatrix} \begin{bmatrix} V_1 - 1 \\ V_2 - 1 \\ \vdots \\ V_n - 1 \end{bmatrix} = - \begin{bmatrix} C_1 & 0 & \cdots & 0 \\ 0 & C_2 & \cdots & 0 \\ \vdots & \vdots & \ddots & \vdots \\ 0 & 0 & \cdots & C_n \end{bmatrix} \begin{bmatrix} \dot{V}_1 \\ \dot{V}_2 \\ \vdots \\ \dot{V}_n \end{bmatrix}$$

(2.8)

$$\text{or} \quad \mathbf{G}(\mathbf{V} - \mathbf{1}) = -\mathbf{C}_{\text{diag}}\dot{\mathbf{V}}$$

$$\text{or} \quad \mathbf{1} - \mathbf{V} = \mathbf{G}^{-1}\mathbf{C}_{\text{diag}}\dot{\mathbf{V}} \qquad (2.9)$$

$$\text{or} \quad \mathbf{1} - \mathbf{V} = \mathbf{RC}_{\text{diag}}\dot{\mathbf{V}}$$

where $\mathbf{R} = \mathbf{G}^{-1}$, and $\mathbf{1} \in \mathbf{R}^n$, with all entries equal to 1.

The Elmore delay at node i is given by

$$\begin{aligned} T_{Di} &= \int_0^\infty [1 - V_i(t)]\, dt \\ &= \int_0^\infty \sum_{k=0}^n R_{i,k} C_k \dot{V}_k\, dt \\ &= \sum_{k=0}^n R_{i,k} C_k \int_0^\infty \dot{V}_k\, dt \\ &= \sum_{k=0}^n R_{i,k} C_k [V_k(\infty) - V_k(0)] \\ &= \sum_{k=0}^n R_{i,k} C_k \end{aligned}$$

$$\mathbf{T_D} = \mathbf{RC}$$

$$= \mathbf{G}^{-1}\mathbf{C}. \qquad (2.10)$$

In general, inverting $\mathbf{G}$ is a tedious and computationally expensive procedure. However, in the case of an RC tree, $\mathbf{G}^{-1}$ can be

2.2. MICROMODELING - THE RC MODEL

written down by inspection as follows.

- Let P_i be the unique path from the input node $\mathbf{n}_0$ to node $\mathbf{n}_i$.
- Let $P_{ij} = P_i \cap P_j$ denote the portion of the path between $\mathbf{n}_0$ and $\mathbf{n}_i$ that is common to the path between $\mathbf{n}_0$ and $\mathbf{n}_j$.
- Let R_{ij} denote the sum of the resistances in P_{ij}; if $P_{ij} = \emptyset$, then $R_{ij} = 0$.

The Elmore delay to node i in the RC tree is then given by the expression

$$T_{Di} = \sum_{j=0}^{n} R_{ij} C_j. \tag{2.11}$$

Example : Consider the circuit shown in Figure 2.3. The node $\mathbf{n}_0$ is the input node. The procedure described above may be used to compute the Elmore delay to various nodes in the circuit. For example,

$$\begin{aligned} T_{Dn_7} =\ & R_1 C_1 + R_1 C_2 + R_1 C_3 + R_1 C_4 + R_1 C_5 + (R_1 + R_6) C_6 \\ & + (R_1 + R_6 + R_7) C_7 + (R_1 + R_6 + R_7) C_8. \end{aligned} \tag{2.12}$$

$$\begin{aligned} T_{Dn_5} =\ & R_1 C_1 + (R_1 + R_2) C_2 + (R_1 + R_2) C_3 \\ & + (R_1 + R_2 + R_4) C_4 + (R_1 + R_2 + R_4 + R_5) C_5 \\ & + R_1 C_6 + R_1 C_7 + R_1 C_8. \end{aligned} \tag{2.13}$$

Several techniques, notably those suggested by Lin and Mead [LM84], Chan and Karplus [CK89], Chan and Schlag [CS89] and

Wyatt [Wya85], may be used for a fast computational approximation to the Elmore delay of a general RC mesh. Other significant work on Elmore delay calculation, including delay estimation of RC networks for slow excitations, and for RC networks where some capacitances are partially charged, is found in [Cha86a, Cha86b].

2.2.5 Penfield-Rubenstein Bounds

Rubenstein, Penfield and Horowitz proposed upper and lower bounds on the step response of an RC tree in [RPH83]. For most practical situations, these bounds are tight, and an average of the upper and lower bounds gives a good estimate of the step response of the tree.

Before proceeding further, we introduce the following time constants

$$T_P = \sum_k R_{kk} C_k \tag{2.14}$$

$$T_{Di} = \sum_k R_{ki} C_k \tag{2.15}$$

$$T_{Ri} = \sum_k \frac{R_{ki}}{R_{ii}} C_k \tag{2.16}$$

where R_{ij} is defined as in Section 2.2.4. Note that the quantity T_{Di} corresponds to the Elmore delay. It can easily be shown from the definitions that

$$0 \leq T_{Ri} \leq T_{Di} \leq T_P. \tag{2.17}$$

2.2. MICROMODELING - THE RC MODEL

The step response bounds are then given by the following equations

$$\underline{v}_i(t) = \begin{cases} 0, & 0 \le t \le T_{Di} - T_{Ri} \\ 1 - \frac{T_{Di}}{t+T_{Ri}}, & T_{Di} - T_{Ri} \le t \le T_P - T_{Ri} \\ 1 - \frac{T_{Di}}{T_P} e^{(T_P - T_{Ri})/T_P} e^{-t/T_P}, & T_P - T_{Ri} \le t < \infty \end{cases} \quad (2.18)$$

$$\overline{v}_i(t) = \begin{cases} 1 - \frac{T_{Di}-t}{T_P}, & 0 \le t \le T_{Di} - T_{Ri} \\ 1 - \frac{T_{Ri}}{T_P} e^{(T_{Di} - T_{Ri})/T_{Ri}} e^{-t/T_{Ri}}, & T_{Di} - T_{Ri} \le t < \infty \end{cases} \quad (2.19)$$

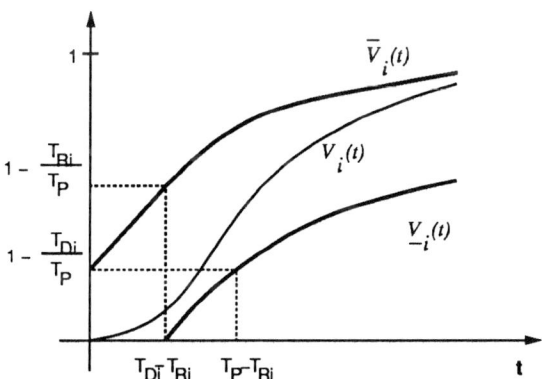

Figure 2.10: The Penfield-Rubenstein bounds.

The general form of these bounds is shown in Figure 2.10. These bounds are tight when $T_{Ri} \simeq T_P$; it can be shown that

$$\overline{v}_i(t) - \underline{v}_i(t) \le \frac{T_P - T_{Ri}}{T_{Ri}}. \quad (2.20)$$

For example, in the case of an RC line (an RC tree without any side branches; for an example, see Figure 2.6(a)), $T_P = T_{Ri}$, and $\underline{v}_i(t) = \overline{v}_i(t)$. While the inequality in Equation (2.20) is conservative, the expression on the right-hand side is a good qualitative indicator of the distance between the bounds.

It may be pointed out that the functions $\underline{v}_i$ and $\overline{v}_i$ are one-to-one functions of t. Hence, bounds on the delay $t(V)$, the time required by the step response to cross a voltage threshold V, can easily be obtained from the above expression.

$$t_{min}(v_i) = \begin{cases} 0, & 0 \leq v_i \leq 1 - \frac{T_{Di}}{T_P} \\ T_{Di} - T_P(1 - v_i), & 1 - \frac{T_{Di}}{T_P} \leq v_i \leq 1 - \frac{T_{Ri}}{T_P} \\ T_{Di} - T_{Ri} + T_{Ri} \cdot \ln\left[\frac{T_{Ri}}{T_P(1-v_i)}\right], & 1 - \frac{T_{Ri}}{T_P} \leq v_i < 1 \end{cases}$$

(2.21)

$$t_{max}(v_i) = \begin{cases} \frac{T_{Di}}{1-v_i} - T_{Ri}, & 0 \leq v_i \leq 1 - \frac{T_{Di}}{T_P} \\ T_P - T_{Ri} + T_P \cdot \ln\left[\frac{T_{Di}}{T_P(1-v_i)}\right], & 1 - \frac{T_{Di}}{T_P} \leq v_i < 1 \end{cases}$$

(2.22)

A more detailed discussion on the tightness of these bounds can be found in [Wya87].

2.3 Macromodeling

Macromodeling [Mat85, Che88] is a technique that was developed to help analyze large circuits. The underlying philosophy is that the information that is of interest to circuit designers is at the gate level, rather than at the transistor level. Hence, individual logic blocks, such as gates, are replaced by equivalent *macromodels*. This serves the purpose of eliminating all nodes internal to a logic gate, thereby reducing the number of nodes and, therefore, the computational complexity. The macromodeling approach is better suited to incorporating information about the shape of the input waveform to a gate, as compared to the RC (micromodeling) approach.

The basic unit in this approach is the *channel-connected component* (henceforth referred to simply as *component*). Each component corresponds to a set of transistors that are connected by drain and source nodes.

More formally, the definition of a component can be given by the following construction: Create an undirected graph, $\mathcal{G}$, with a vertex for each circuit node and an edge between the drain and source nodes of each transistor. Next, split the vertices corresponding to the ground node, the supply (V_{DD}) node, and the primary input nodes, such that each of these vertices is incident on only one edge after splitting, to obtain a new graph, $\mathcal{G}'$. A component is then a set of transistors corresponding to the edges within a connected com-

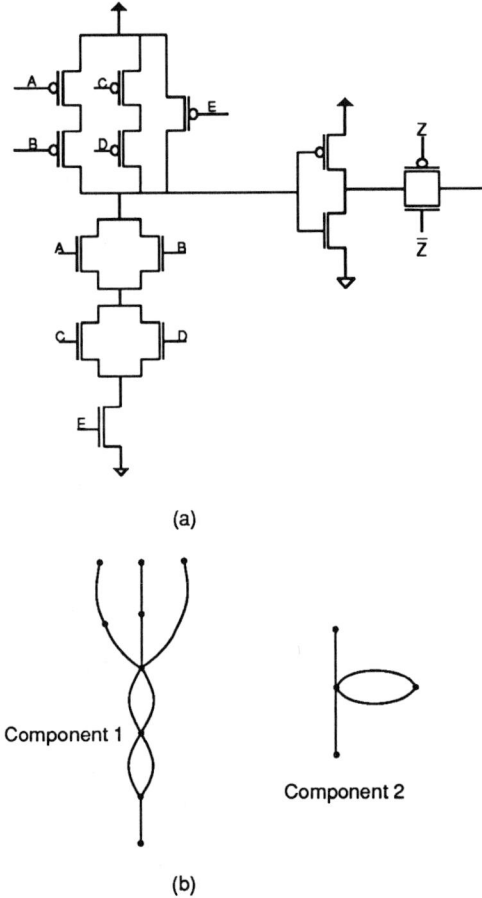

Figure 2.11: Identification of channel-connected components.

ponent of the graph, $\mathcal{G}'$. This process is illustrated by an example in Figure 2.11. The graph $\mathcal{G}'$ corresponding to the circuit in Figure 2.11(a) is shown in Figure 2.11(b). This graph has two strongly connected components, which correspond to the channel-connected components.

The input nodes of a component consist of all of the gate nodes

2.3. MACROMODELING

of transistors in the component, and any drain or source node of a transistor in the component that is also a primary input. A component's output nodes include any drain or source node of a transistor in the component.

The macromodeling method classifies components according to whether they are static or dynamic. For each class of components, a basic element has to be identified. A technique for mapping any component in that class to the basic element is then devised; the delay of the component is taken to be the delay of the equivalent basic element. Clearly, the accuracy of macromodeling of a component depends on of the accuracy of the mapping function.

For example, in case of static CMOS gates, except composite gates such as that shown in Figure 2.1, the basic element could be the inverter, and all other gates may be mapped to an inverter according to some mapping function. Two different approaches may be used to calculate inverter delay times as a function of various parameters, such as the n-transistor width, W_n, the p-transistor width, W_p, the load capacitance, C_L, and the input slew rate S_I.

2.3.1 The Table Look-up Method

In this procedure, a table of the inverter delay as a function of various parameters, such as W_p, W_n, C_L and S_I, is maintained. Note that since the table makes some presumptions about the technology parameters, it has to be generated again each time the technology changes. Such an approach allows for fast delay computation; however, it suffers from the fact that the size of the table could grow very rapidly. For example, for a table with four parameters, and 10 data points for each parameter, the number of entries would be 10^4. Various techniques are invoked to reduce this number. For instance, the effects of W_n and W_p could be combined into a single parameter $\beta = W_n/W_p$. Another simplification could be attributed to the experimental observation that the delay time is approximately linearly shifted by a certain amount that is a function of the input wave shape for reasonable input waveforms; the effect of input slew rate S_I could be empirically quantified as a functional relationship.

Rao *et al.* [RTH83, Rao85] have presented an approach for delay calculation for MOS circuits using the table look-up method. Although the method described here works for NMOS circuits and not CMOS circuits, it is instructive to examine it as an illustration of the table look-up technique.

Five primitives are defined, and for each primitive, two types of inputs are possible. The five primitives, shown in Figure 2.12 are

2.3. MACROMODELING

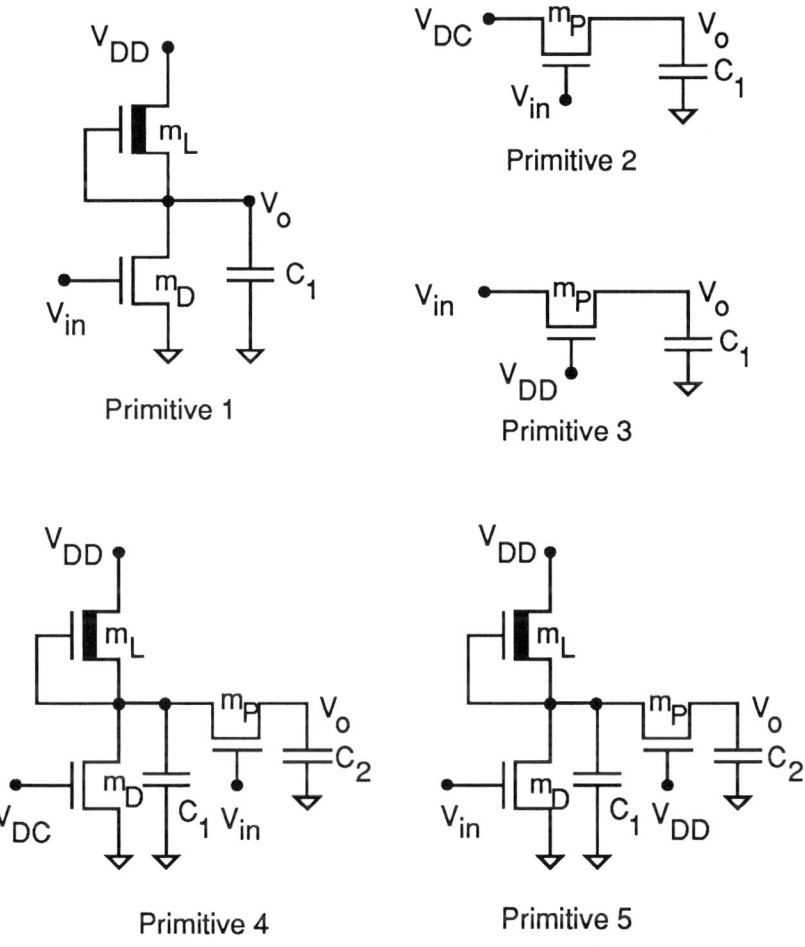

Figure 2.12: The five basic primitives for Rao et al.'s approach.

[1] A simple inverter driving a lumped grounded capacitance, C_1. The input V_{in} is the signal at the gate node of the driver, and the output is the voltage across the capacitor.

Type 0 : V_{in} rising from 0V to 5V.

Type 1 : V_{in} falling from 5V to 0V.

[2] A pass transistor whose drain node is connected to a constant dc voltage source, and the gate node is driven by a rising pulse from 0V to 5V. The source node, which is the output, is connected to a lumped capacitor C_2.

Type 0 : dc source, V_{DC}, at 0V.

Type 1 : dc source, V_{DC}, at 5V.

[3] A pass transistor whose drain node is connected to a voltage source V_{in}, and the gate node is held fixed at 5V. The output node is the source node connected to a lumped capacitor C_2.

Type 0 : V_{in} rising from 0V to 5V.

Type 1 : V_{in} falling from 5V to 0V.

[4] A simple inverter driving a pass transistor. A lumped capacitance C_1 is connected to the output of the inverter, and a grounded capacitance C_2 is connected to the source node of the pass transistor, which is also the output node for this configuration. A rising signal from 0V to 5V is applied at the gate node of the pass transistor, while the gate of the driver tran-

2.3. MACROMODELING

sistor of the inverter is held fixed.

Type 0 : gate of driver, V_{DC}, at 0V.

Type 1 : gate of driver, V_{DC}, at 5V.

[5] The same configuration as in [4], except that the gate of the pass transistor is held fixed at 5V, while the input signal is applied to the gate of the transistor.

Type 0 : V_{in} rising from 0V to 5V.

Type 1 : V_{in} falling from 5V to 0V.

In each of the above primitives, the input waveform V_{in} varies between 0 and $V_{DD} = 5V$. For a fixed input waveform, the shape of the output V_o could depend upon several circuit device and process parameters. The parameters considered are the following:

- zero backgate bias threshold voltages (VTOs) for both enhancement and depletion mode devices
- the resistance of the device, which is a function of the ratio of its channel length, L, to its width, W
- the transconductance parameter, $K'_p = \mu_n \epsilon_{ox}/t_{ox}$, which is a function of the carrier mobility, μ_n, the permittivity of the oxide material, ϵ_{ox}, and the thickness of the oxide, t_{ox}
- the capacitance at each node.

Among these parameters, it is assumed that all enhancement (depletion) transistors have the same zero backgate bias threshold voltage, VTO_E (VTO_D), and that these values remain fixed for a given technology (for example, $VTO_E = 1.0V$ and $VTO_D = -3.0V$). The rest of the parameters are allowed to vary between the different devices and nodes in the network. For the five primitives described above, let R_D, R_L and R_P denote the device resistances of the driver transistor, m_D, the load transistor, m_L, and the pass transistor, m_P, respectively. Standard driver, load and pass transistors are chosen, with R_{DS}, R_{LS} and R_{PS} being the resistances of these standard devices, respectively. A typical set of standard devices is

Load : W/L = 0.5

Driver : W/L = 2.0

Pass Transistor : W/L = 1.0

For the above choice of standard load and driver devices, we notice that $R_{LS}/R_{DS} = 4$; this is referred to as the standard inverter ratio, δ_S.

Let C_{is} denote the standard capacitance in the ith primitive. Typically, C_{is} is 0.01 pF for i = 1,2,3, and 0.1 pF for i = 4,5. A primitive is a standard primitive if

$R_D = R_{DS}$, $R_L = R_{LS}$, and $C_1 = C_{1S}$ in primitive 1.

$R_P = R_{PS}$, $C_2 = C_{2S}(C_{3S})$ in primitive 2 (primitive 3).

$R_P = R_{PS}$, $C_2 = C_{4S}(C_{5S})$ in primitive 4 (primitive 5).

2.3. MACROMODELING

In primitives 4 and 5, two dimensionless quantities, $\beta = R_D/R_P$, and $\gamma = C_1/C_2$, are defined, and are used to completely specify the standard primitive. β and γ are allowed to vary over the ranges $[\beta_{min}, \beta_{max}]$ and $[\gamma_{min}, \gamma_{max}]$, respectively.

Consider one of the above primitives. V_{in} is treated as an analog ramp waveform with a full swing of V_{DD}. Let the times at which the waveform crosses the two threshold voltages, V_L and V_H, be t_1 and t_2, respectively. As a result of the change in the input, the output waveform crosses the threshold at times t'_1 and t'_2, respectively. Define $\Delta_{in} = t_2 - t_1$, a measure of the slew rate of the input signal, and two delay quantities, $\Delta t_a = t'_1 - t_1$, the *inertial delay*, and $\Delta t_b = t'_2 - t_2$, the *rise/fall delay*. Given t_1 and the two delay quantities, t'_1 and t'_2 can easily be computed. We will use the symbol Δt_o to refer to the two delay quantities, Δt_a and Δt_b, collectively.

Consider the problem of computing Δt_o for standard primitives. First consider the standard primitive 1 with rising inputs, i.e., type 1, driven by an input ramp V_{in} with a certain value of Δ_{in}. The circuit is then simulated using an accurate circuit simulator, such as SPICE2 [Nag75], which calculates the falling output waveform, V_o. From the V_{in} and V_o waveforms, the threshold crossing times, t_1, t_2, t'_1, and t'_2, are calculated, from which the delays, Δt_a and Δt_b, are found. This is repeated for a falling input ramp, i.e., type 1, with the same slew rate as before, and two more delay values are computed.

This experiment is then repeated with input ramps of different slew rates, producing a delay table with four delay values (namely, Δt_1 and Δt_2, for type 0 and type 1) for each value of Δ_{in}. The procedure is repeated to generate delay tables for the standard primitives 2 and 3. The tables in all three cases are one-dimensional, since Δ_{in} is the only variable.

For primitives 4 and 5, C_1, R_D and R_P have to be specified to describe the circuit completely. This is done with the help of parameters β and γ. For fixed values of these parameters, we obtain $C_1 = \gamma C_2$, and $R_D = \beta R_P$, where R_P and C_2 take on standard values. By varying Δ_{in}, β and γ, three-dimensional delay tables are generated, with each entry containing four delay values as before.

Nonstandard components are mapped on to these standard primitives using a technique described in [Rao85], and the table generated by the techniques described above is used to compute their delay.

2.3.2 The Analytical Method

This method expresses the inverter delay in terms of an analytic expression that is a a function of various parameters. This procedure is more rigorous than the table look-up method. However, as device geometries continue to shrink, second-order effects need to be considered, which could make it difficult or impossible to produce a suitable closed-form expression. As a result, the table look-up

2.3. MACROMODELING

method is likely to become more prevalent in the future, particularly with the advent of smaller feature sizes.

As an example to illustrate the analytical method of macromodeling, the delay calculation technique used in [Che88] for static CMOS gates (excluding composite gates of the type in Figure 2.1) is described here. The primitive gate considered is the inverter, and an outline of the procedure by which the delay of an inverter is calculated and a description of the process of mapping a gate to the primitive gate follow.

Delay Model for the Inverter

Figure 2.13(a) is an inverter circuit with capacitive load C_L. Assume that the input is a positive going signal, starting at time 0, with rise time, t_r, and slope, $m = \frac{V_{DD}}{t_r}$, as shown in the Figure 2.13(b). The typical output waveform of the inverter is shown in Figure 2.13(c).

The delay time is the time when the output voltage is half of the power supply V_{DD}, and is calculated in two stages. The first stage is for output voltage dropping from V_{DD} to $0.9V_{DD}$ and is denoted as t_1; the second stage is for output voltage dropping from $0.9V_{DD}$ to $0.5V_{DD}$ and is denoted as t_2. The total delay time

$$t_d = t_1 + t_2. \tag{2.23}$$

In the first stage, the pFET functions in the linear region and the

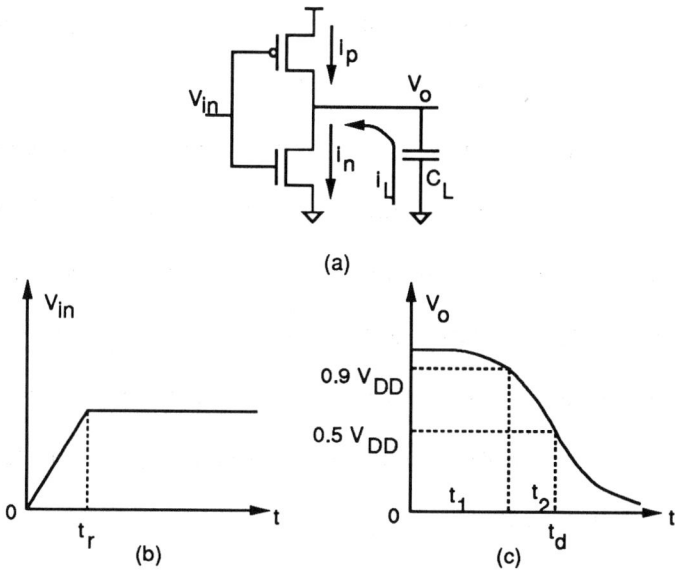

Figure 2.13: (a) Inverter circuit (b) Input signal (c) Output waveform.

nFET in the saturation region. The current discharging the load capacitance C_L can be written as

$$i_L = i_n - i_p = -C_L \frac{dV_o}{dt}. \qquad (2.24)$$

The currents i_n and i_p can be expressed separately as

$$i_n = \frac{\beta_{In}}{2}(V_{gs} - V_{Tno})^2 = \frac{\beta_{In}}{2}(mt - V_{Tno})^2. \qquad (2.25)$$

$$i_p = \beta_{Ip}\left[(V_{gs} - V_{Tpo})V_{ds} - V_{ds}^2/2\right] \qquad (2.26)$$

$$= \beta_{Ip}\left[(mt - V_{DD} - V_{Tpo}) \cdot (V_o(t) - V_{DD}) \right.$$

$$\left. -(V_o(t) - V_{DD})^2/2\right]. \qquad (2.27)$$

2.3. MACROMODELING

With the calculation confined to the first region from V_{DD} to $0.9V_{DD}$, $V_o(t)$ can be approximated with a small error by averaging these two values, i.e., $V_o = \frac{V_{DD}+0.9V_{DD}}{2}$. Equation (2.27) can then be recast to

$$i_p = \beta_{Ip}\left[(V_{DD} + V_{Tpo} - mt)\left(\frac{V_{DD}}{20}\right) - \frac{1}{2}\left(\frac{V_{DD}}{20}\right)^2\right]. \quad (2.28)$$

Substitute Equation (2.25) and (2.28) into Equation (2.24) to obtain

$$i_p - i_n = \beta_{Ip}\left[(V_{DD} + V_{Tpo})\left(\frac{V_{DD}}{20}\right) - \frac{1}{2}\left(\frac{V_{DD}}{20}\right)^2 \right.$$

$$\left. - m\left(\frac{V_{DD}}{20}\right)t\right] - \frac{\beta_{In}}{2}(mt - V_{Tno})^2$$

$$= C_L \frac{dV_o(t)}{dt}. \quad (2.29)$$

Equation (2.29) can be manipulated to give

$$m^2 t^2 + \left[\frac{2\beta_{Ip}}{\beta_{In}}\left(\frac{V_{DD}}{20}\right)m - 2mV_{Tno}\right]t + V_{Tno}^2$$

$$- \frac{2\beta_{Ip}}{\beta_{In}}\left[(V_{DD} + V_{Tpo})\left(\frac{V_{DD}}{20}\right) - \frac{1}{2}\left(\frac{V_{DD}}{20}\right)^2\right] = -\frac{2C_L}{\beta_{In}}\frac{dV_o}{dt}. \quad (2.30)$$

or, more concisely,

$$m^2 t^2 + K_1 t + K_2 = -\frac{2C_L}{\beta_{In}}\frac{dV_o}{dt}. \quad (2.31)$$

where

$$K_1 = \frac{\beta_{Ip} m V_{DD}}{10\beta_{In}} - 2mV_{Tno}.$$

$$K_2 = V_{Tno}^2 - \frac{2\beta_{Ip}}{\beta_{In}}\left[(V_{DD} + V_{Tpo})\left(\frac{V_{DD}}{20}\right) - \frac{1}{2}\left(\frac{V_{DD}}{20}\right)^2\right] \quad (2.32)$$

Integrating both sides of Equation (2.31) for V_o from V_{DD} to $0.9V_{DD}$ over the time period t_1,

$$\int_0^{t_1}(m^2t^2 + K_1t + K_2)dt = -\int_{V_{DD}}^{0.9V_{DD}}\frac{2C_L}{\beta_{In}}dV_o. \quad (2.33)$$

The evaluation of Equation (2.33) yields the following cubic equation

$$t_1^3 + \frac{3K_1}{2m^2}t_1^2 + \frac{3K_2}{m^2}t_1 - \frac{3C_LV_{DD}}{5\beta_{In}m^2} = 0. \quad (2.34)$$

where t_1 is solved by using the cubic root formula [Bey78]. Depending on whether t_r is greater than t_1, there can be two cases.

<u>Case A : $t_r < t_1$</u>

The input signal rises to V_{DD} before the output voltage drops to $0.9V_{DD}$; the output voltage V_X at time point $t = t_r$ can be calculated as shown in the following equation.

$$\int_0^{t_r}(m^2t^2 + K_1t + K_2)dt = -\int_{V_{DD}}^{V_x}\frac{2C_L}{\beta_{In}}dV_o. \quad (2.35)$$

$$V_x = V_{DD} - \frac{\beta_{In}}{2C_L}\left(\frac{m^2t_r^3}{3} + \frac{K_1t_r^2}{2} + K_2t_r\right). \quad (2.36)$$

Beyond the time point t_r, the input signal is in its full swing and the pFET is essentially shut off. All of the current is provided by the

2.3. MACROMODELING

nFET, which is in saturation. The time taken in bringing the output voltage down to $0.5V_{DD}$ from V_X is simply

$$t' = \frac{C_L(V_x - 0.5V_{DD})}{I_{nc}} = \frac{0.5V_{DD}C_L - \frac{\beta_{In}}{2}\left(\frac{m^2 t_r^3}{3} + \frac{K_1 t_r^2}{2} + K_2 t_r\right)}{I_{nc}} \quad (2.37)$$

where I_{nc} is the discharging current flowing through the nFET which is now in saturation. This current can be expressed as

$$I_{nc} = \frac{\beta_{In}}{2}(V_{DD} - V_{Tno})^2. \quad (2.38)$$

The total delay time is the summation of t_r and t' or $t_d = t_r + t'$. The slope of the output waveform, $V_o(t)$, can be represented by the slope at the point when $V_o(t) = 0.5V_{DD}$, and is expressed as $|dV_o/dt| = I_{nc}/C_L$.

Case B : $t_r \geq t_1$

The input signal is still in its positive-going stage, with a slope of m. In the second region where the output voltage is dropping from $0.9V_{DD}$ to $0.5V_{DD}$, the nFET is strongly on while the pFET is barely on. The current flowing through the nFET is essentially equal to the discharge current i_L. The state equation of Equation (2.24) becomes

$$i_n = \frac{\beta_{In}}{2}(V_{gs} - V_{Tno})^2 = \frac{\beta_{In}}{2}(mt_1 + mt - V_{Tno})^2 = -C_L\frac{dV_o}{dt}.. \quad (2.39)$$

Let $A = mt_1 - V_{Tno}$ and integrate both sides of Equation (2.39) for

V_o from $0.9V_{DD}$ to $0.5V_{DD}$ over the time period t_2.

$$\int_0^{t_2} (m^2t^2 + 2Amt + A^2)dt = -\int_{0.9V_{DD}}^{0.5V_{DD}} \frac{2C_L}{\beta_{In}} \frac{dV_o}{dt}. \qquad (2.40)$$

The following cubic equation is obtained.

$$t_2^3 + \frac{3A}{m}t_2^2 + \frac{3A^2}{m^2}t_2 = \frac{2.4C_LV_{DD}}{\beta_{In}m^2}. \qquad (2.41)$$

The time t_2 can be solved using a cubic root formula [Bey78].

Again, depending on whether $t_1 + t_2$ is greater than t_r, we can have the following two cases.

Case B.1 : $t_1 + t_1 \leq t_r$

The output voltage drops to half of V_{DD} while the input signal is still in its positive-going stage. The delay time is the summation of both stages or $t_d = t_1 + t_2$. The slope of the output voltage, $V_o(t)$, at time point t_d can be obtained from Equation (2.39) and can be expressed as $\mid dV_o/dt \mid = \beta_{In}/2C_L(mt_d - V_{Tno})^2$.

Case B.2 : $t_1 + t_2 > t_r > t_1$

The input voltage rises to V_{DD} before the output voltage dropped to $0.5V_{DD}$. The output voltage V_Y at time point t_r can be calculated from the following equation.

$$\int_0^{t_r-t_1} (m^2t^2 + 2Amt + A^2)dt = -\int_{0.9V_{DD}}^{V_Y} \frac{2C_L}{\beta_{In}} dV_o. \qquad (2.42)$$

2.3. MACROMODELING

Let $\hat{t} = t_r - t_1$. Voltage V_Y can be expressed as

$$V_Y = 0.9V_{DD} - \frac{\beta_{In}}{2C_L}\left(\frac{m^2\hat{t}^3}{3} + Am\hat{t}^2 + A^2\hat{t}\right). \tag{2.43}$$

Beyond time point t_r, the input signal is in its full swing and all the current is provided by the nFET, which is in saturation. The time taken in bringing V_o down to $0.5V_{DD}$ from V_Y can be expressed as

$$\begin{aligned}\tilde{t} &= \frac{(V_Y - 0.5V_{DD})C_L}{I_{nc}} \\ &= \frac{0.4V_{DD}C_L - \frac{\beta_{In}}{2}(\frac{m^2\hat{t}^3}{3} + Am\hat{t}^2 + A^2\hat{t})}{I_{nc}}\end{aligned} \tag{2.44}$$

where I_{nc} is the current already calculated in Equation (2.38). The delay time is therefore $t_d = t_1 + \hat{t} + \tilde{t}$ or $t_d = t_r + \tilde{t}$. The slope of output voltage, $V_o(t)$, is equal to $|dV_o/dt| = I_{nc}/C_L$.

The flowchart for the inverter delay time calculation is shown in Figure 2.14.

If the input is a negative-going signal with fall time t_f and slope $m = |\frac{V_{DD}}{t_f}|$, the inverter will operate in a way dual to the case in which the input signal is positive going. The derivation can just follow the procedures in the previous sections and the delay model is readily available with only the following parameter substitutions.

$$\begin{aligned}\beta_{In} &\leftrightarrow \beta_{Ip} \\ t_r &\leftrightarrow t_f \\ V_{Tno} &\leftrightarrow -V_{Tpo}\end{aligned} \tag{2.45}$$

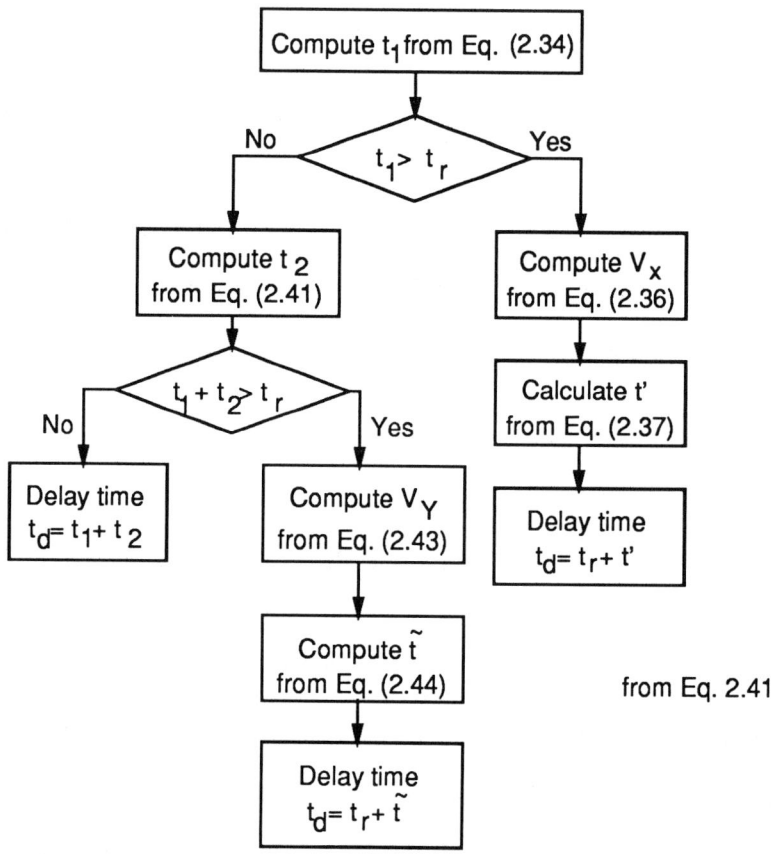

Figure 2.14: Flowchart for inverter delay time calculation.

Delay Models for Other Primitive Gates

The delay models for primitive gates other than the inverter gate can be obtained by mapping them to their equivalent inverter gates. The equivalent channel width can be determined by using the worst-case analysis under the assumption that channel lengths are the same.

2.3. MACROMODELING

1. If transistors are connected in series, the equivalent channel width is

$$\frac{1}{W_{eq}} = \frac{1}{\frac{1}{W_1} + \frac{1}{W_2} + \ldots + \frac{1}{W_n}}. \qquad (2.46)$$

2. If transistors are connected in parallel, the equivalent channel width is

$$W_{eq} = \min(W_1, W_2, \ldots, W_n). \qquad (2.47)$$

In this parallel connection case, the worst case analysis assumes that more than one transistor cannot be on at the same time.

A two-input NAND circuit example is shown in Figure 2.15(a) and its equivalent inverter is shown in Figure 2.15(b). The transis-

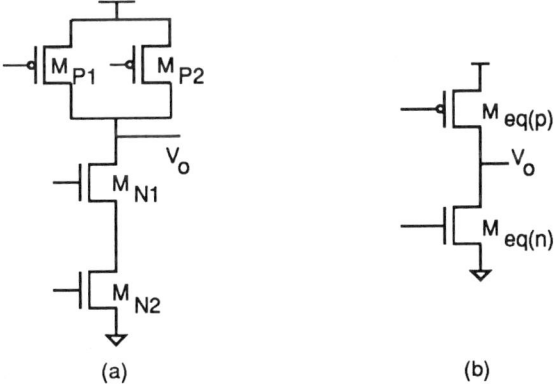

Figure 2.15: (a) Two-input NAND gate (b) The equivalent inverter.

tor $M_{eq}(P)$ has an equivalent p-channel width $W_{eq}(P)$ equal to the smaller one of W_{p1} and W_{p2} and transistor $M_{eq}(N)$ has an equivalent n-channel width $W_{eq}(N) = \frac{W_{n1}W_{n2}}{W_{n1}+W_{n2}}$.

More details about the analytical approach for other gates, including domino CMOS gates, are provided in [KCh90, Che88].

2.4 Worst-case Delay Estimation

The worst-case delay estimation at the output of a component consists of finding the latest rising and falling transitions at the output node, over all allowable input transitions. The waveform at the input nodes of a component could be a steady logic 0, a steady logic 1, a logic 0 to logic 1 transition, or a logic 1 to logic 0 transition.

Alternatively stated, the problem is to find the set of transistors that are on during the worst-case transition. For a CMOS gate, an n-transistor will remain *on* if the signal at its gate is at logic 1, or is in transition from logic 0 to logic 1; it is *off* if it has a logic 0 signal at its gate. Similarly, a p-transistor will remain on if the signal at its gate is at logic 0, or is in transition from logic 1 to logic 0; it will remain off if the signal is at logic 1.

The set of transistors thus found forms a unique path between the output node and the V_{DD} (ground) node, whenever each path between the output and a given supply node is independent of other paths, i.e., when there is no signal that feeds the gate nodes on two different paths in the pullup (pulldown) network. To justify this, suppose to the contrary that there were two paths P_1 and P_2 between the supply and the output nodes, as in Figure 2.16. Then the gate

2.4. WORST-CASE DELAY ESTIMATION

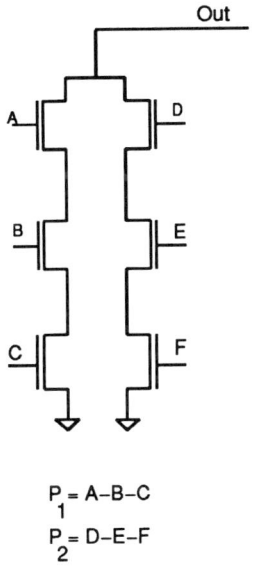

$P_1 = $ A–B–C
$P_2 = $ D–E–F

Figure 2.16: An example with two paths between output and ground.

delay due to transistors on P_1 alone or on P_2 alone would be less than the delay for the case in which all transistors on $P_1 \cup P_2$ are on. In the case in which two or more paths between an output node and a given supply node are dependent, it is possible for more than one path to ground to be activated. However, considering only a unique path between the supply and output nodes would only be pessimistic in estimating the worst-case delay.

The slowest paths for rising and falling output transitions can be determined by enumerating all paths between the output node and V_{DD} or ground, respectively, in the component. The number of paths could grow exponentially, which could be a serious problem, particularly when computationally expensive transistor models are used, as in the case of CRYSTAL [Ous85] and TAMIA [Dag87], or when the delay computation has to be carried out a large number of times, as in the case of transistor size optimization. Certain pruning procedures, such as those in [SVR91, DGR92], may be used to improve the computational speed. The procedures used in [SVR91] are described in Section 2.7.

2.5 Delay Calculation at the Circuit Level

At the circuit level, the delay calculation deals primarily with purely combinational logic circuits. Any sequential circuit can be decomposed into combinational logic blocks that lie between latches. Since each of these logic blocks has to satisfy a particular timing requirement dictated by system clock signals, it is the delay of a combinational logic block that has practical significance.

In this section, we first look at how the combinational subnetworks are extracted from a sequential circuit, and then examine the application of the PERT method to finding the delay of a combinational subnetwork.

2.5.1 Combinational Subnetwork Extraction

A procedure for identifying feedback latches and transmission gate latches in a sequential circuit is described in this section.

2.5. DELAY CALCULATION AT THE CIRCUIT LEVEL

The circuit is represented by a graph, G, with vertices corresponding to components, and with edges drawn from a component to other components to which it fans out. Feedback loops in the circuit (e.g., cross-coupled NAND gates), which manifest themselves as strongly connected components in this graph, are identified using Tarjan's algorithm [Eve79].

Next, each clock signal is traced from the primary inputs, proceeding from a component to the components to which it fans out, until the signal intersects either a feedback loop or the gate node of a transistor in a transmission gate. Such a feedback loop or transmission gate is identified as a latch. Thus, this procedure identifies latches which are clocked not only by clock signals at the primary input, but also by qualified clock signals.

To illustrate how transparent transmission-gate latches may be handled, we examine the procedure used in TILOS [DFH89]. TILOS assigns segment numbers according to the following algorithm: when data is already setup at the input to a transmission gate latch before the clock arrives to let it flow through, the segment number of the transmission gate latch for that data polarity is assigned to be 1. The output of a combinational logic gate is assigned the segment number of its critical input. If a signal arrives at the input of a transmission gate latch after the clock has turned it on, then the segment number of the output node of the transmission gate is one greater than the segment number of the input node; i.e., the signal flows through the transmission gate, and is then on the next segment of its timed path. Although in principle, a signal can travel through any number of latches during their transparent period, in practical designs, the number of latch-to-latch segments allowed is limited by a user-settable parameter, MAXSEG, that defaults to two. If a signal path has MAXSEG latch-to-latch segments, and does not arrive at the final latch before it becomes transparent, the circuit is considered to have failed to satisfy its timing specifications.

All latches are then removed from the circuit. For transmission gate latches, this could result in a single component being broken up into two or more components. A new graph $\hat{G}$ is formed, in the same way as G, to represent this new circuit. A breadth-first search [Eve79] of $\hat{G}$ can detect strongly connected components of this new circuit; each such strongly connected component corresponds to a combinational subcircuit that lies between a set of input latches and a set of output latches. The clock arrival times at these latches are used to determine the timing requirements of the combinational subnetworks.

Before proceeding further, we define the following quantities

- The *set-up time*, T_{su}, is the time interval required for the input to be stable before it is clocked into a latch.
- The *hold time*, T_h, is the time interval after the clock transition

2.5. DELAY CALCULATION AT THE CIRCUIT LEVEL

has occurred, during which the input must be held stable.

Given the triggered times for the input latch, T_{il}, and the output latch, T_{ol}, a technique described in [LKT87] can be adapted to determine the timing requirement between this input/output pair for the combinational subnetwork.

Let $T_{PD,in}$ be the propagation delay of the input latch. If $T_X = T_{il} + T_{PD,in} > T_{ol}$, then we increment T_{ol} by $Tperiod_{ol}$ until $T_X < T_{ol}$. This value of T_{ol} is the transition time at the output latch that must be satisfied by the critical (longest) path, with consideration for set-up time requirements.

As illustrated in Figure 2.17, the critical path delay $T_{d,max}$, triggered by the transition at time T_{il}, must be such that the output transition occurs by time $T_1 = T_{ol} - T_{su}$, i.e.,

$$T_{d,max} < T_{ol} - T_{su} - T_{il}. \qquad (2.48)$$

Similarly, the hold time requirements dictate conditions that the shortest path delay, $T_{d,min}$ must satisfy. For edge-triggered latches, the output transition caused by the clock transition at time $T_{il} + Tperiod_{il}$ must not occur earlier than $T_{ol} + T_h$, i.e.,

$$T_{d,min} > T_{ol} + T_h - (T_{il} + Tperiod_{il}). \qquad (2.49)$$

When a signal streams through a latch in midphase, then the resulting transition at the output can occur no earlier than $T_{il} + T_{PD,in} +$

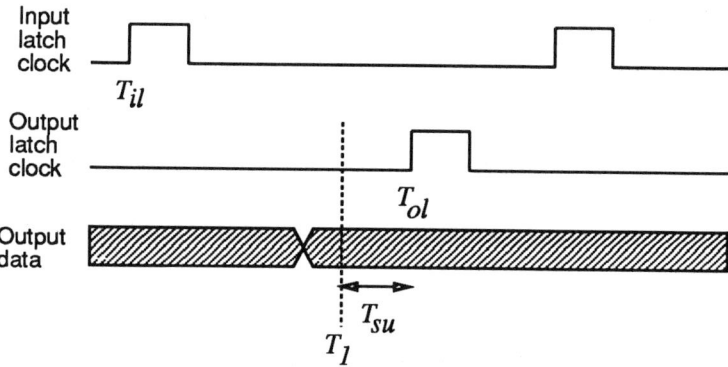

Figure 2.17: Set-up time requirements.

$T_{d,min}$. This must occur at a time later than $T_{ol}+T_h$. In other words, $T_{d,min}$ must satisfy

$$T_{d,min} > T_{ol} + T_h - (T_{il} + T_{PD,in}). \qquad (2.50)$$

2.5.2 The PERT Method

A well-known statistical scheduling method, PERT (Program Evaluation and Review Technique), is useful in finding the delay of a circuit, given the delays of individual gates. The method, first applied by Kirkpatrick and Clark [KC66] in 1966 (where it was illustrated by an example of a ferrite core system design), is now an integral part of many fast algorithms for circuit delay calculation.

The procedure is best illustrated by means of a simple example. Consider the circuit in Figure 2.18. Each box represents a channel-connected component (defined in Section 2.3). The number within

2.5. DELAY CALCULATION AT THE CIRCUIT LEVEL 57

the box represents the delay associated with it. We assume that the worst-case arrival time for a transition at any primary input, i.e., at the inputs to boxes A, B, C, and D, is 0.

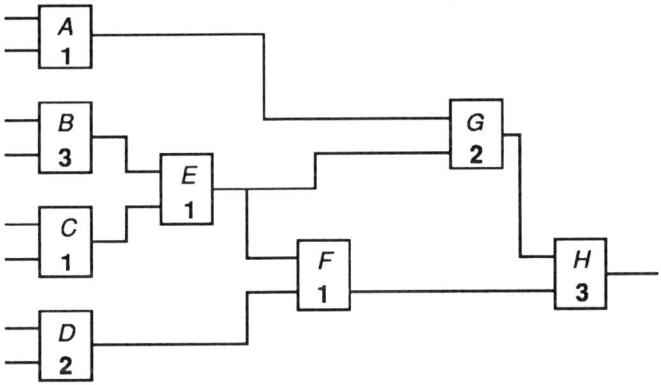

Figure 2.18: The PERT technique.

A component is said to be *ready for processing* when the signal arrival time information is available for all of its inputs. Initially, since signal arrival times are known only at the primary inputs, only those components that are fed solely by primary inputs are ready for processing. In the example, these components would be A, B, C, and D. These are placed in a queue and are scheduled for processing.

In the iterative process, the component at the head of the queue is scheduled for processing. Each processing step consists of

- Finding the latest arriving input to the component, which triggers the output transition. This involves finding the maximum of all worst-case arrival times of inputs to the component.

- Adding the delay of the component to the latest arriving input time, to obtain the worst-case transition time at the output.
- Checking all of the components that the current component fans out to, to find out whether they are ready for processing. If so, the component is added to the tail of the queue.

The iterations end when the queue is empty.

In the example, the algorithm is executed as follows:

Step 1. In the initial step, A, B, C, and D are placed in the queue.

Step 2. A is scheduled. The latest input transition for A is 0, the delay of A is 1; hence, the latest output transition for A is at time $(0+1) = 1$. No additional elements can be added to the queue.

Step 3. B is scheduled. The latest output transition is found to be at time $(0+3) = 3$. No additional elements can be placed in the queue.

Step 4. C is scheduled, and the worst-case output transition occurs at time $(0+1) = 1$. At this point, all input information for component E is available, and it is placed at the tail of the queue.

Step 5. D is scheduled; the output transition time is found to be $(0+2) = 2$.

Step 6. E is scheduled; the latest arriving input comes in at time 3; hence, the worst-case output transition occurs at time $(3+1) = 4$. At this point, F and G are added to the tail of the queue.

Step 7. F is scheduled; its latest output transition is at time $(4+1)$

2.5. DELAY CALCULATION AT THE CIRCUIT LEVEL 59

= 5.

Step 8. G is scheduled; its latest output transition occurs at time (4+2) = 6. H is added to the tail of the queue.

Step 9. H is scheduled; its worst-case output transition is at time (6+3) = 9. The queue is now empty and the algorithm terminates.

The worst-case delays at the output of each component are shown in Figure 2.19.

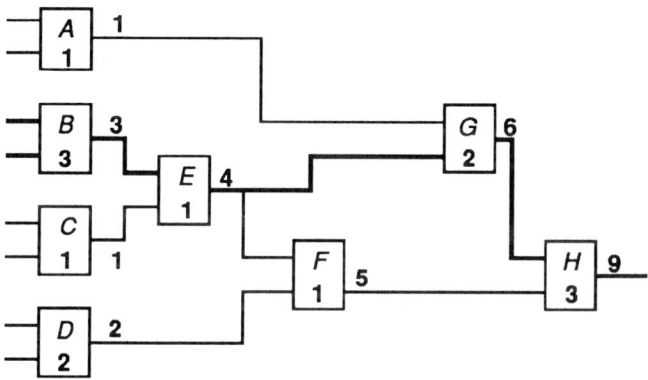

Figure 2.19: Circuit delays obtained from the PERT technique.

The *critical path*, defined as the path between an input and an output with the maximum delay, can now easily be found by using a traceback method. We begin with the component whose output is the primary output with the latest transition time. This is the last component on the critical path. Next, the latest arriving input to this component is identified. The component that causes this transition is the preceding component on the critical path. The process is repeated recursively until a primary input is reached.

In the example, we begin with component H at the output. The latest arriving input to this component is caused by component G, which thus precedes H on the critical path. Similarly, the transition at G is caused by the latest arriving input, which comes in from component E, and so on. By continuing this process, the critical path from the input to the output is identified as B-E-G-H.

In the case of CMOS circuits, the rise and fall delay transitions are calculated separately. For inverting CMOS gates, the latest arriving input rise (fall) transition triggers off a fall (rise) transition at the output. This can easily be incorporated into the PERT method described above, by maintaining two numbers, t_r and t_f, for each gate, corresponding to the worst-case rise (high transition) and fall (low transition) delays from a primary input. To obtain the value of t_f at an output, the largest value of t_r at an input node is added to the worst-case fall transition time of the component; the computation of t_r at an output is analogous. For noninverting gates, t_r and t_f are obtained by adding the rise (fall) transition time to the worst-case input rise (fall) transition time. A more detailed description is provided in Section 2.7.

2.6 Posynomial Delay Modeling

In certain optimization applications, it is useful to express the circuit delay in the form of a certain class of functions known as *posynomials*.

2.6. POSYNOMIAL DELAY MODELING

A *posynomial* is a function g of a positive variable $\mathbf{x} \in \mathbf{R}^n$ that has the form

$$g(\mathbf{x}) = \sum_j \gamma_j \prod_{i=1}^n x_i^{\alpha_{ij}} \qquad (2.51)$$

where the exponents $\alpha_{ij} \in \mathbf{R}$ and the coefficients $\gamma_j > 0$. Roughly speaking, a posynomial is a function that is similar to a polynomial, except that

- The coefficients α_{ij} must be positive.
- An exponent α_{ij} could be any real number, and not necessarily a positive integer, unlike the case of polynomials.

A posynomial has the useful property that it can be mapped onto a convex function through an elementary variable transformation, $(x_i) = (e^{z_i})$ [Eck80]. Such a functional form is very desirable, since convex functions have certain properties which can be exploited while performing optimization.

A treatment of posynomial delay models for digital CMOS circuits is provided in [HKE89].

As an example to illustrate the usefulness of posynomials, consider the problem of transistor sizing. Fishburn and Dunlop[FD85] first showed that the delay along a path of the circuit may be expressed as a posynomial function of the channel widths of transistors.

They formulated the transistor sizing problem as

$$\text{minimize} \quad \sum_i x_i \quad (2.52)$$

$$\text{subject to} \quad Circuit\ Delay \leq T_{spec} \quad (2.53)$$

where x_i is the size (channel width) of the ith transistor in the circuit. The objective function is clearly a posynomial function of the x_i's. Note that the circuit delay constraint is equivalent to specifying that each path delay must be less than T_{spec}. Thus, if the path delays can be described by posynomials, both the objective function and the constraints for this optimization problem are posynomials. The transformation $(x_i) = (e^{z_i})$ maps the transistor sizing problem to that of minimizing a convex function over convex constraints. This type of optimization problem is known as a convex programming problem, and has the useful feature that any local minimum is also a global minimum.

2.7 A Case Study: iCONTRAST's Timing Analyzer

In this section, we present the timing analyzer used in the transistor size optimization algorithm in [SVR91]. For this application, a posynomial form is chosen for the delay function, since it has some advantages, mentioned in Section 2.6, that are useful for optimization purposes. Since the size of every transistor is a design variable in this problem, this approach has to work at the transistor level. Macromodeling at the gate level cannot be used here since we wish to operate at the transistor level; therefore, micromodeling techniques are employed.

2.7.1 Transistor-level Micromodeling

A MOS transistor is modeled as a voltage-controlled switch with an on-resistance, R_{on}, between drain and source, and three grounded capacitances, C_d, C_s, and C_g, at the drain, source, and gate terminals, respectively, as shown earlier in Figure 2.5. The resistance and capacitances associated with a MOS transistor of channel width x are taken to have the following dependence [SFD88] on x :

$$R_{on} \propto 1/x \qquad (2.54)$$

$$C_d, C_s, C_g \propto x. \qquad (2.55)$$

The exact capacitance relations for an n-transistor of width x can be understood with the help of Figure 2.20. (Note that the expressions associated with a p-transistor are analogous.) As illustrated in the figure, the parameter L stands for the length of the channel, and d_s and d_d refer to the lengths that the diffusion extend beyond the channel area on the source and drain ends, respectively.

$$C_d = C_{JA,n} \cdot d_d \cdot x + 2 \cdot C_{JP,n} \cdot (d_d + x) \qquad (2.56)$$

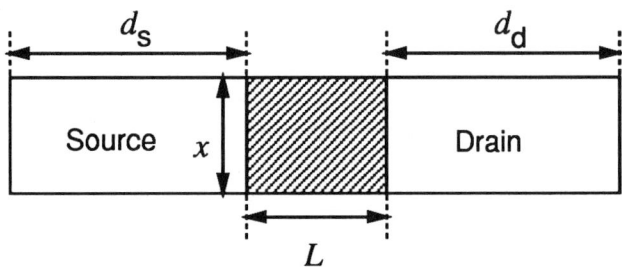

Figure 2.20: The physical parameters associated with a transistor.

$$C_s = C_{JA,n} \cdot d_s \cdot x + 2 \cdot C_{JP,n} \cdot (d_s + x) \quad (2.57)$$

$$C_g = C_{GTA} \cdot L \cdot x + 2 \cdot C_{GTP} \cdot (L + x) \quad (2.58)$$

where
- C_{JA} : Diffusion area capacitance (pF/μm^2)
- C_{JP} : Diffusion perimeter capacitance (pF/μm)
- C_{GTA} : Gate area capacitance (pF/μm^2)
- C_{GTP} : Gate perimeter capacitance (pF/μm).

The on-resistance of an n-transistor (p-transistor) is given by the expression $R_{on,n} = R_n/x$ ($R_{on,p} = R_p/x$). The coefficients R_n and R_p are calibrated using expressions for the rise and fall times, respectively, of a CMOS inverter driving a load capacitance.

Consider the inverter shown in Figure 2.21 (a). We use the above model to derive an equivalent RC circuit; the equivalent RC circuit for the fall transition is shown in Figure 2.21 (b). The delay of the fall transition at the output of this inverter (derived in Section 6.2.4)

A CASE STUDY: iCONTRAST'S TIMING ANALYZER

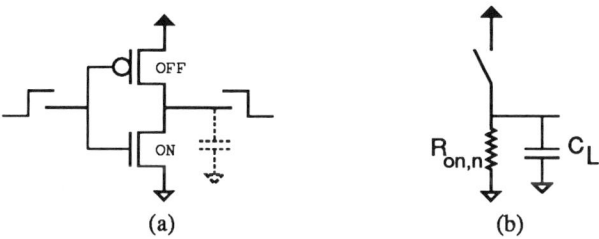

Figure 2.21: (a) An inverter undergoing a fall transition (b) Its equivalent RC network.

is given by

$$t_f = \frac{L_n}{K_n(V_{DD} - V_{Tn})} \left[\ln \left[3 - 4\frac{V_{Tn}}{V_{DD}} \right] + \frac{2V_{Tn}}{V_{DD} - V_{Tn}} \right] \frac{C_L}{x}. \quad (2.59)$$

Note that this equation is equivalent to Equation (6.8).

The Elmore delay of the equivalent RC circuit is given by

$$t_{f,Elmore} = \frac{R_n}{x} C_L. \quad (2.60)$$

Comparing equations (2.59) and (2.60), we find that calibrating R_n on the inverter gives us the expression

$$R_n = \frac{L_n}{K_n(V_{DD} - V_{Tn})} \left[\ln \left[3 - 4\frac{V_{Tn}}{V_{DD}} \right] + \frac{2V_{Tn}}{V_{DD} - V_{Tn}} \right]. \quad (2.61)$$

In a similar manner, the value of R_p, when calibrated on the rise transition of the CMOS inverter, is given by the expression

$$R_p = \frac{L_p}{K_p(V_{DD} - |V_{Tp}|)} \left[\ln \left[3 - 4\frac{|V_{Tp}|}{V_{DD}} \right] + \frac{2|V_{Tp}|}{V_{DD} - |V_{Tp}|} \right].$$

$$(2.62)$$

2.7.2 The Delay Estimation Algorithm

In this section, an algorithm for estimating the worst-case delay through the circuit, over all possible input combinations, is described.

Consider a combinational CMOS circuit with a set of primary input nodes and primary output nodes. The circuit is first divided into components; recall from Section 2.3 that each component corresponds to a set of transistors that are connected by drain and source nodes.

The PERT method described in Section 2.5 is used to compute the maximum overall rise and fall delays between the primary inputs and primary outputs of the circuit, and the *critical path*, i.e., the path between an input and an output with the maximum delay. For each output node of each component of the circuit, we assign the numbers

- t_r, the total rise delay from the primary inputs to that node.
- t_f, the total fall delay from the primary inputs to that node.
- Δ_r, the Elmore delay of an RC network that corresponds to the worst-case rise scenario.
- Δ_f, the Elmore delay of an RC network that corresponds to the worst-case fall scenario.

The output transition waveform is modeled as a function that varies linearly with time. The transition times of the rising and falling

A CASE STUDY: iCONTRAST'S TIMING ANALYZER 67

waveforms at the output of the component are taken to be $2\Delta_r$ and $2\Delta_f$, respectively. Figure 2.22 illustrates the definition of t_f and Δ_f; t_r and Δ_r are defined in a similar manner.

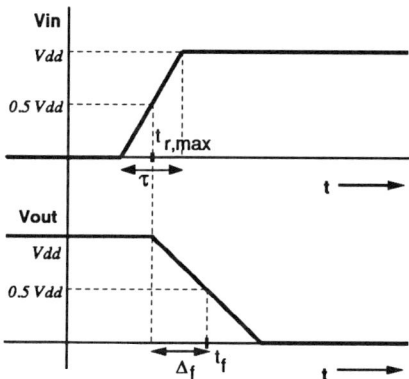

Figure 2.22: Graphical illustration of t_f and Δ_f.

Finding the Worst-case Elmore Delay

Once a component has been scheduled for evaluation by PERT, the worst-case Elmore delay at an output node of the component (as described in Section 2.4) must be found. Let *o* denote an output node of the component. The algorithm for finding the worst-case fall delay at *o* is described below; the worst-case rise delay at *o* can be found in an analogous manner.

The component is represented by an undirected weighted graph, G, with an edge between the drain and source nodes of each transistor in the component. Edge weights are given by the resistance R_{on} of the corresponding transistor. The V_{DD} node and all of its incident edges are then removed from the graph. Let $t_{r,max}$ denote the maximum value of t_r among all input nodes of the component, and suppose this occurs at the gate node of an n-type transistor corresponding to an edge e_{max} in G. It is assumed that the worst-case path is the *largest resistive path* (LRP), i.e., the path of largest weight, between o and ground that passes through e_{max}. This assumption is valid when the load capacitance at the output node is much greater than the internal capacitance at any node that lies on any path between the output node and the ground node through e_{max}, as is often the case in CMOS circuits.

For example, if we consider the fall transition for Component 1 in Figure 2.11, if input D is the latest arriving rising input to the component, then the edge e_{max} corresponds to the n-transistor whose gate is connected to D. We then find the LRP through edge e_{max}, and consider that to be the worst-case path between output and ground.

Since finding the LRP is equivalent to the longest path problem in a graph which is NP-hard [Eve79], the following heuristic can be used to perform this task. This heuristic is exact for series-parallel graphs, such as CMOS complex gates, and can be outlined as follows.

LRP Algorithm

T = maximum weighted spanning tree in G containing e_{max} such that the path P between o and ground in T contains e_{max}
$maxW$ = sum of weights of edges in P
$LINK$ = edges in $G - T$

```
for each edge e ∈ LINK {
    T₁ = T ∪ e
    P₁ = max weight o-to-ground path in T₁ through emax
    W = sum of weights of edges in P₁
    if ( W > maxW ) {
        e' = any edge in P - (P ∩ P₁)
        T = T₁ - e', P = P₁, maxW = W
    }
}
```

The heuristic begins by finding a maximum weighted spanning tree T of G that contains the edge e_{max}, using a variant of Prim's algorithm [Eve79]. Let P' denote the *unique* path in T between o and ground. If P' contains e_{max}, set P to P'; otherwise, an edge, $e \notin T$, is added to T such that $T + e$ has a path P between o and ground through e_{max}, and the e is the edge of greatest weight among all edges that satisfy this condition. The introduction of e creates a unique cycle; an edge e', such that $e' \in P'$ and $e' \notin P$, is removed from $T + e$, to give a new initial tree T.

The edges that are not in T constitute the set of *links*. A link is then added to the present tree T to produce a subgraph T_1 that contains a *unique cycle*. Therefore, there can be at most two paths from o to ground in T_1. The path of larger weight is called P_1. If the weight of P_1 is larger than that of P, then the present tree T is updated by removing any edge from T_1 that belonged to P but not to P_1. Also, P is reset to P_1 and the heuristic proceeds to process the next link, and so on, until all links of the original tree have been processed. The path between o and ground in the final tree produced by the heuristic is considered to be the LRP. In the case of series-parallel graphs, the heuristic does indeed generate the path of largest resistance from output to ground; in other cases (such as graphs with bridges), it gives a good approximation.

Now, consider any spanning tree T_w of the graph G. If P_p and P_q are the paths to ground from nodes p and q, respectively, in T_w, let R_{pq} denote the resistance of the path $P_p \cap P_q$. The Elmore delay, as discussed in Section 2.2.4, between o and the ground node in the RC-tree represented by T_w is given by

$$\sum_{j \in T_w} R_{oj} C_j \qquad (2.63)$$

where C_j is the capacitance to ground at node j in T_w. Note that while finding the Elmore delay, the capacitances which lie between the switching transistor and the supply rail are assumed to be at the voltage level of the supply rail at the time of the switching transition, and do not contribute to the Elmore delay.

In order to find a tree that contains the LRP and which maximizes the Elmore delay, certain edges must be added to the LRP in such a way that R_{oj} is maximized for every node j in the graph. The algorithm to construct the worst-case tree T_w from the LRP is as follows. Initially, T_w is taken to be the LRP itself. For a node

$n_1 \notin \mathcal{T}_w$ the algorithm finds a node $n_2 \in \mathcal{T}_w$ that is farthest from the ground node and is connected to n_1 by a path that does not intersect $\mathcal{T}_w$. This path is then added to $\mathcal{T}_w$ and the procedure is repeated until all nodes of G are included in the tree $\mathcal{T}_w$. The worst-case fall delay at o is then computed using (2.63).

Example

Consider the graph G shown in Figure 2.23. Assume that the LRP between the output node o and ground has been found to be **d,e** . Initially, $\mathcal{T}_w$ is taken to be the LRP **d,e** . Consider node n_1 which is connected to node o through several paths, one of which is **j,k** . This path is added to $\mathcal{T}_w$ which now becomes **d,e,j,k** . It is worth pointing out here that the exact branches that are chosen, which lie off the LRP, are immaterial to the Elmore delay at node o, since their resistance values do not enter the Elmore delay expression; hence, in the last step, we could have chosen either **h,k** or **i,k** or **j,k** or **h,l** or **i,l** or **j,l** for inclusion into the tree. Note that both nodes n_1 and n_2 are now part of the tree $\mathcal{T}_w$. The nodes n_4 and n_5 are then added to the tree by adding the edges **a** and **b** , respectively. Finally, the node n_6 is added to the tree by adding the edge **f** to it. This completes the formation of the worst-case tree which is **d,e,j,k,a,b,f** indicated by the bold edges in Figure 2.23.

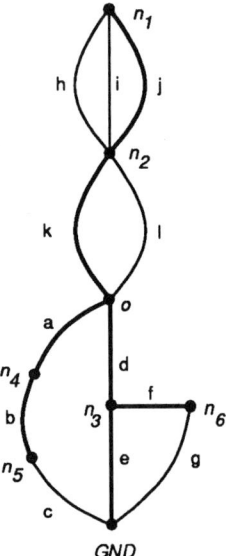

Figure 2.23: Extending the LRP to a tree.

Finally, the value of t_f for output node o is computed by adding $t_{r,max}$, the Elmore delay of the worst-case RC network, Δ_f, and a term related to the transition time of the rising input at the input node corresponding to the worst-case Elmore fall delay [HJ87].

A more detailed description of how the effect of input transition time is incorporated is provided later in this section. This procedure is repeated for all output nodes of the component.

It is worth pointing out here that in reality, the input that actually triggers off the worst-case transition is not necessarily the latest arriving input as we have assumed, since the delay depends not only on the arrival times, but also on the Elmore delay and the input transition time. However, such an approximation is appropriate for the transistor sizing problem for most real circuits (with special handling for primary inputs), since for most components in a circuit, the input arrival time is the largest contributor to the output delay value. If an incorrect critical input is chosen in any iteration, the transistor sizing algorithm attempts to reduce the delay along the LRP corresponding to that input, so that in some subsequent iteration, the correct critical input will be the latest arriving input. Moreover, our approximations are borne out by SPICE comparisons shown in Section 2.7.3.

The value of t_r, the worst-case rise delay at each output node of the component, can be found in a similar manner. The weighted graph representing the component is constructed as before except that the ground node is removed instead of the V_{DD} node. The rest of the procedure to find the worst-case Elmore rise delay is identical to that of the fall delay except that the role of the ground node is replaced by the V_{DD} node, and the roles of t_r and Δ_r are exchanged with those of t_f and Δ_f in the fall delay case.

In other delay estimators, the Elmore rise and fall delays are computed directly from the LRP without appending additional edges to extend it to the worst-case RC tree as described above, and the delays calculated thus fail to be the worst-case delays.

Delay Model for Components under Nonstep Inputs

In [HJ87], it has been shown that a good approximation to the delay, Δ, of a CMOS inverter under excitation from a nonstep input of the form

$$v_{in} = \begin{cases} 0 & t < 0 \\ \frac{t}{\tau} \cdot V_{DD}, & 0 < t < \tau \\ V_{DD}, & t > \tau \end{cases} \quad (2.64)$$

is given by

$$\Delta = \Delta_{step} + \frac{\tau}{6}\left[1 + \frac{2V_{Tn}}{V_{DD}}\right] \quad (2.65)$$

where
$\quad V_{Tn}\ $ = threshold voltage of nMOS transistor
$\quad V_{DD}\ $ = supply voltage
$\quad \Delta_{step}$ = inverter transition delay under a step input excitation

Δ is defined as the difference between the time when the output signal crosses the $V_{DD}/2$ level and the time at which the input signal reaches $V_{DD}/2$. The falling output signal can be approximated by using a form similar to the input waveform v_{in} in (2.64). The relationship between the input and output signals in our model, for the falling output transition, is shown in Figure 2.22.

A general complex gate such as the AOI gate, when excited by a step excitation, may be replaced by an equivalent inverter I whose size is determined by the Elmore delay of the worst-case RC tree described earlier. For an excitation of the type in (2.64), we may consider the general complex gate as being equivalent to the inverter I being excited by the same excitation. Hence, (2.65) can also be used for complex gates.

The form of the path delay under step excitations is described in [FD85]. We examine the change required in this form to include the effect of waveforms with nonstep transitions as described in (2.64), under the assumption that the signal at the output of a component is modeled as described earlier in this section.

Let $\Delta_{i,step}$ refer to the delay of component i on a path of the circuit, with all input waveforms having step transitions. The delay of the circuit, $Delay_{step}$, is given by

$$Delay_{step} = \Delta_{1,step} + \Delta_{2,step} + \cdots + \Delta_{n,step}. \quad (2.66)$$

When we incorporate the effect of the transition time, and adopt a simplifying assumption that the magnitude of the threshold voltage is the same for nMOS and pMOS enhancement mode transistors, the delay along the path is given by

$$\begin{aligned}
Delay_n &= \Delta_{1,step} + [\Delta_{2,step} + \alpha \cdot Delay_1] \\
&\quad + [\Delta_{3,step} + \alpha \cdot (Delay_2 - Delay_1)] \\
&\quad + [\Delta_{4,step} + \alpha \cdot (Delay_3 - Delay_2)] + \cdots \\
&\quad + [\Delta_{n,step} + \alpha \cdot (Delay_{n-1} - Delay_{n-2})] \\
&= \Delta_{1,step} + \Delta_{2,step} + \cdots + \Delta_{n-1,step} + \Delta_{n,step} + \alpha \cdot Delay_{n-1}
\end{aligned}$$

$$= w_1 \cdot \Delta_{1,step} + w_2 \cdot \Delta_{2,step} + \cdots + w_n \cdot \Delta_{n,step}. \qquad (2.67)$$

where

$Delay_k$ = circuit delay up to k^{th} component from primary inputs

$$\alpha = \frac{1}{3} \cdot \left[1 + \frac{2\,|\,V_T\,|}{V_{DD}}\right]$$

V_T = threshold voltage (assumed equal in magnitude for nMOS, pMOS for simplicity)

$$w_k = \sum_{i=0}^{n-k} \alpha^i.$$

Thus, the delay is expressed by the weighted sum of the $\Delta_{i,step}$ values. Since each of the $\Delta_{i,step}$ expressions is posynomial [FD85], and the w_i's are constant, the expression for delay along a path under excitations with nonstep transitions is a posynomial.

In the case in which the threshold voltage, V_T, is different for n- and p-type transistors, the form of (2.67) remains the same, but the expression for each w_k is more involved.

2.7.3 A Comparison with SPICE

Results showing the performance of iCONTRAST's timing analyzer on several circuits, in comparison with SPICE values, are presented in this section. To show the improvement provided by iCONTRAST, a comparison is shown with an Elmore delay based timing analyzer. This Elmore delay based timing analyzer is essentially iCONTRAST's timing analyzer without the enhancements which enable it to take into consideration the input waveform rise time, and the effect of capacitances that do not lie on the LRP.

The circuits include a chain of eight inverters (Inv8), a complete binary tree of seven two-input NAND gates (Tree), and a two-bit adder using complex gates (Add2). Each data set for a circuit corresponds to a different set of transistor sizes in the circuit. For the circuit Inv8 and Tree, the iCONTRAST delay values, shown by the (a) lines, are in excellent agreement. For a more complicated circuit such as Add2, with complex gates and more transistors, the accuracy deteriorates, but is still within reasonable limits. In all cases, there is a substantial improvement over the accuracy of the unenhanced version shown by the (b) lines. These results are displayed graphically in Figure 2.24.

2.8 Summary

In this chapter, various approaches for fast estimation of circuit delay have been presented. Two basic approaches to delay modeling, micromodeling and macromodeling, are studied. In the micromodeling approach, individual transistors are modeled, usually by resistors and capacitors, so that each gate is represented by an RC network. The Elmore time constant approach and the Penfield-Rubenstein bounds for finding the delay through an RC network are reviewed. In the case of macromodeling, a higher level of abstraction, such as the gate level, is employed. In the case of gate-level macromodeling, a basic primitive gate (usually an inverter) is chosen. Each gate is mapped on to an equivalent primitive gate. The delay of the primitive gate is then found using table look-ups, or by plugging in the values of various parameters into an analytical formula for the basic primitive. At the circuit level, the PERT method for circuit-level delay estimation is studied.

Finally, after a brief motivation for the use of posynomial delay estimators, a case study of the iCONTRAST timing analyzer, which uses posynomial models for delay estimation, is examined in detail. This timing analyzer incorporates the effects of waveform slopes, and calculates the worst-case circuit delay. Comparisons with SPICE show that the enhancements in this approach over previous RC approaches afford a considerable improvement in the accuracy of the estimated worst-case delay.

2.8. SUMMARY

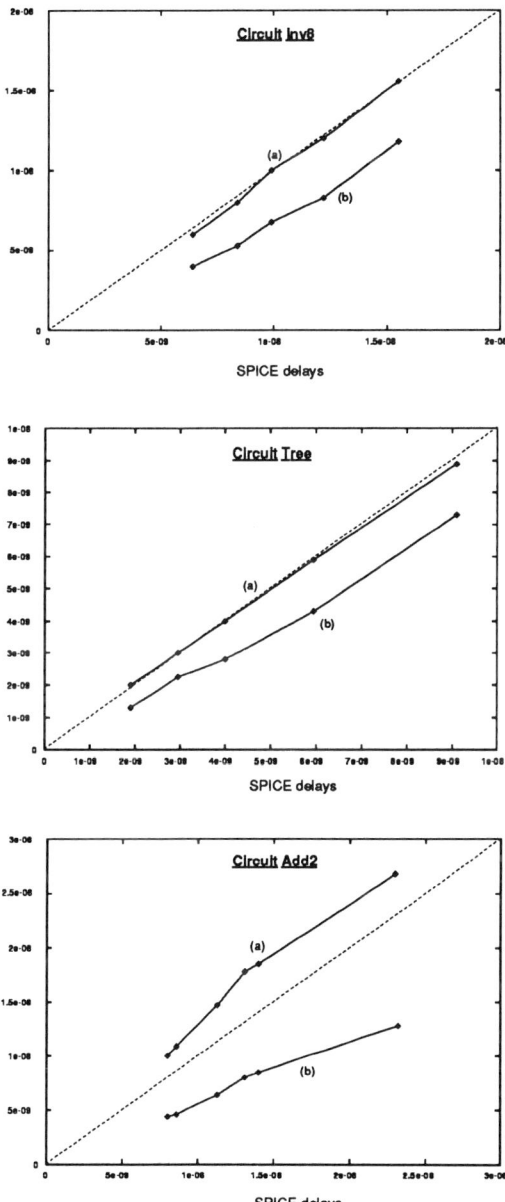

Figure 2.24: (a) iCONTRAST's timing analyzer vs. SPICE (b) An Elmore delay based timing analyzer without iCONTRAST's enhancements vs. SPICE

Chapter 3

Transistor Sizing Algorithms : Existing Approaches

3.1 Introduction

Circuit delays in integrated circuits often have to be reduced to obtain faster response times. A typical digital integrated circuit consists of multiple stages of combinational logic blocks that lie between latches that are clocked by system clock signals. For such a circuit, delay reduction must ensure that valid signals are produced at each output latch of a combinational block, before any transition in the signal clocking the latch. In other words, the worst-case input-output delay of each combinational stage must be restricted to be below a certain specification.

Given the circuit topology, the delay of a combinational circuit can be controlled by varying the sizes of transistors in the circuit. Here, the size of a transistor is measured in terms of its channel width, since the channel lengths of MOS transistors in a digital circuit are generally uniform. In any case, what really matters is the ratio of channel width to channel length, and if channel lengths are not uniform, this ratio can be considered as the size. In coarse terms, the circuit delay can usually be reduced by increasing the sizes of certain transistors in the circuit (the actual relation between transistor sizes and the circuit delay is more complex, as illustrated in the following example). Hence, making the circuit faster usually entails the penalty of increased circuit area. The area-delay trade-off involved here is, in essence, the problem of transistor size optimization.

Example

A simple example from [Hed87] serves to illustrate how the delay varies with transistor sizes. Consider the chain of three CMOS inverters shown in Figure 3.1(a). For simplicity, assume that all of the transistors have the same channel width. Let the width of both the n-type and p-type transistors in gate 2 be w_2, and let D be the total delay through the three gates.

Consider the effect of increasing w_2, while keeping the size of

3.1. INTRODUCTION

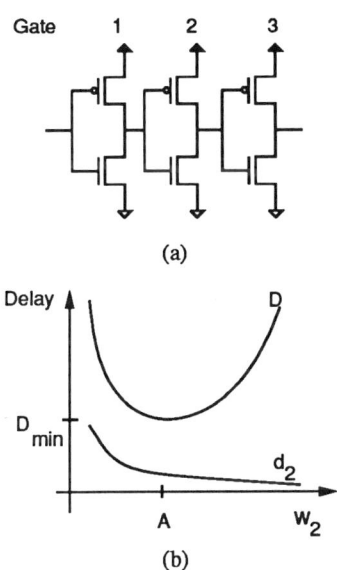

Figure 3.1: (a) A chain of three inverters (b) Effect of transistor sizes on delay for the three-inverter chain.

the transistors in gates 1 and 3 fixed. This causes the magnitude of the output current of gate 2 to increase. Thus the time required, d_2, for gate 2 to drive its output signal will decrease monotonically (Figure 3.1(b)). However, increasing w_2 also increases the capacitive load on the output of gate 1, thus slowing down the output transition of the first gate. Beyond a certain point $w_2 = A$, the total delay, D, starts to increase with respect to w_2, which shows the nonmonotonicity of the delay-area relationship.

Problem Definition

For a combinational circuit, the transistor sizing problem is most commonly formulated as

$$\begin{aligned} & minimize \quad Area \\ & subject\ to \quad Delay \leq T_{spec}. \end{aligned} \quad (3.1)$$

Note that this formulation is directly related to the sequential circuit scenario where each combinational subcircuit must satisfy certain delay specifications with minimal area overhead, and is hence the most useful formulation, in practice.

In addition to Eq. (3.1), several other area-delay formulations have been suggested. These include

- minimize $Area \cdot Delay$
- minimize $Delay$ subject to $Area \leq A_{spec}$.

In addition to these, certain formulations that include power dissipation as an objective, or in terms of constraints, have also been suggested.

In this chapter, we present a broad (but not exhaustive) overview of various transistor sizing algorithms that have been proposed in the literature.

Since many algorithms use delay models that are similar to that used by TILOS, we describe the TILOS model in some detail in Section 3.2.2. For other algorithms, however, our discussion deals primarily with the optimization technique used; unless otherwise specified, the timing model used by these methods is similar to that used by TILOS.

3.2 The TILOS Algorithm

The algorithm that was implemented in TILOS (TImed LOgic Synthesizer) [FD85, DFH89, HSF89] was the first to make use of the fact that the area and delay can be represented as posynomial functions of the transistor sizes. In this section, we describe the details of this algorithm.

3.2.1 The Area Model

The area of a circuit cannot easily be represented as a function of transistor sizes, since finding the area occupied by a circuit, given a circuit-level description (rather than an actual layout), is a combinatorial problem. This is unfortunate, since a closed functional form facilitates the application of optimization techniques. As an approximation, the following formula is used by many transistor sizing algorithms, to estimate the active circuit area.

$$Area = \sum_{i=1}^{n} x_i \qquad (3.2)$$

where x_i is the size of the ith transistor and n is the number of transistors in the circuit. In other words, the area is approximated as the sum of the sizes of transistors in the circuit which, from the definition Eq. (2.51), is clearly a posynomial function of the x_i's.

3.2.2 The TILOS Delay Model

We examine delay modeling in TILOS at the transistor, component and circuit levels.

Transistor Level Model

At the transistor level, the relations in Eq. (2.55) are used to model individual transistors.

Component Level Model

At the component level, TILOS operates in the following manner. For each transistor in a pullup or pulldown network of a complex gate, the largest resistive path from the transistor to the gate output is computed, as well as the largest resistive

3.2. THE TILOS ALGORITHM

path from the transistor to a supply rail. Thus, for each transistor, the network is transformed into an equivalent RC line corresponding to this path (as against the RC tree that is used by the iCONTRAST algorithm in Section 2.7, which may contain resistors that do not lie on the largest resistive path), and the Elmore time constant for this RC line is computed. This Elmore delay corresponds to the delay of the gate when the transition is caused by the transistor under consideration. The greatest Elmore delay thus calculated is taken as the worst-case delay of the gate.

Circuit Level Model

At the circuit level, the PERT technique, described in Section 2.5, is used to find the circuit delay. Since each gate delay is a posynomial, and the circuit delay found by the PERT technique is a sum of gate delays, the circuit delay is also a posynomial function of the transistor sizes.

Recall that while finding the Elmore delay, the capacitances which lie between the switching transistor and the supply rail are assumed to be at the voltage level of the supply rail at the time of the switching transition, and do not contribute to the Elmore delay. For the example in Figure 3.2, the capacitance at node n_1 is ignored while computing the Elmore delay, the expression for which is

$$(R_1 + R_2)C_2 + (R_1 + R_2 + R_3)C_3. \qquad (3.3)$$

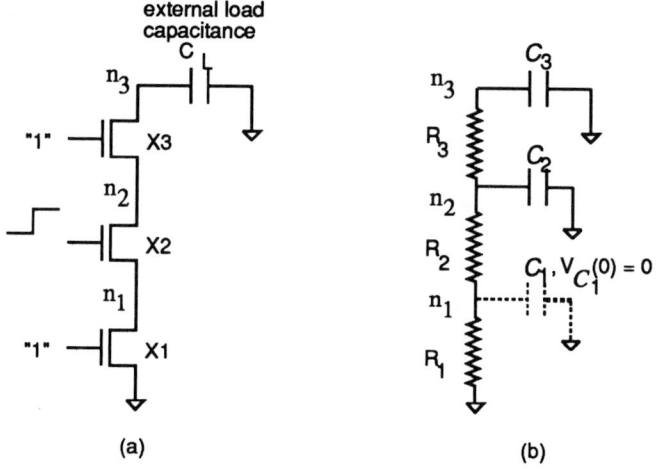

Figure 3.2: (a) A sample pulldown network (b) Its RC representation.

Each R_i is inversely proportional to the corresponding transistor size, x_i, and each C_i is some constant (for wire capacitance) plus a term proportional to the width of each transistor whose gate, drain or source is connected to node i. Thus, Eq. (3.3) can be rewritten as

$$(A/x_1 + A/x_2)(Bx_2 + Cx_3 + D) + (A/x_1 + A/x_2 + A/x_3)(Bx_3 + E)$$

which is a posynomial. In general, the path delay can be written as

$$\sum_{i,j=1}^{n} a_{ij}\frac{x_i}{x_j} + \sum_{i=1}^{n} \frac{b_i}{x_i} + K. \qquad (3.4)$$

3.2.3 The TILOS Optimizer

The optimization algorithm used by TILOS assumes an initial solution where all transistors are at the minimum allowable size. For a general sequential circuit, TILOS first extracts the combinational subnetworks and their input-output timing requirements, and then performs the steps described in this section on each combinational subnetwork. Keeping this in mind, for the rest of this section we can assume, without loss of generality, that the circuit to be optimized is purely combinational.

In each iteration, a static timing analysis is performed on the circuit, which assigns two numbers to each electrical node: t_f, the latest fall transition time, and t_r, the latest rise transition time. This timing analysis is used to determine the critical path for the circuit. Let N be the primary output node on the critical path. The algorithm then walks backward along the critical path, starting from N. Whenever an output node of a component, $Comp_i$, is visited, TILOS examines the largest resistive path between V_{DD} and the output node (if $Comp_i$'s t_r causes the timing failure at N) or the largest resistive path between ground and the output node (if $Comp_i$'s t_f causes the timing failure at N). This includes

- The *critical transistor*, i.e., the transistor whose gate is on the critical path. In Figure 3.2, X2 is the critical transistor.

- The *supporting transistors*, i.e., transistors along the largest resistive path from the critical transistor to the power supply (V_{DD} or ground). In Figure 3.2, X1 is a supporting transistor.
- The *blocking transistors*, i.e., transistors along the highest resistance path from the critical transistor to the logic gate output. In Figure 3.2, X3 is a blocking transistor.

Using the expression for the circuit delay given by Eq. (3.4), TILOS finds the sensitivity, which is the reduction in circuit delay per increment of transistor size, for each critical, blocking and supporting transistor. The procedure of sensitivity computation is treated in greater detail in Section 3.2.4. The size of the transistor with the greatest sensitivity is increased by multiplying it by a constant, BUMPSIZE, a user-settable parameter that defaults to 1.5.

The above process is repeated until

- all constraints are met, implying that a solution is found, or
- the minimum delay state has been passed, and any increase in transistor sizes would make it slower instead of faster, in which case TILOS cannot find a solution.

Note that since in each iteration, exactly one transistor size is changed, the timing analysis method can employ incremental simulation techniques to update delay information from the previous

3.2. THE TILOS ALGORITHM

iteration. This substantially reduces the amount of time spent by the algorithm in critical path detection.

3.2.4 Sensitivity Computation

Figure 3.3 illustrates a configuration in which the critical path extends back along the gate of the upper transistor, which is the critical transistor.

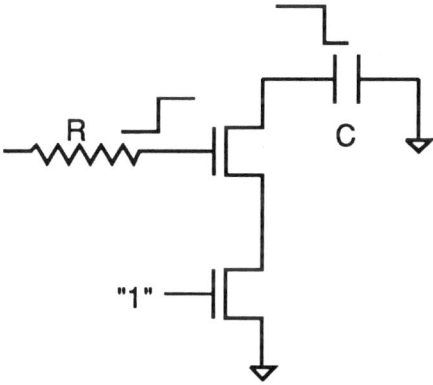

Figure 3.3: Sensitivity calculation in TILOS.

The sensitivity for this transistor is calculated as follows: set all transistor sizes, except x, to the size of the critical transistor. R is the total resistance of an RC chain driving the gate and C is the total capacitance of an RC chain being driven by the configuration. The total delay, $D(x)$, of the critical path is

$$D(x) = K + RC_u x + \frac{R_u C}{x} \qquad (3.5)$$

where R_u and C_u are the resistance and capacitance of a unit-sized transistor, K is a constant that depends on the resistance of the bottom transistor, the capacitance in the driving RC chain and the resistance in the driven RC chain. The sensitivity, $D'(x)$, is then

$$D'(x) = RC_u - \frac{R_u C}{x^2}. \tag{3.6}$$

When $D'(x)$ is set to zero, the value x_0 that minimizes the delay is

$$x_0 = \sqrt{\frac{R_u C}{C_u R}} = \text{constant} \cdot \sqrt{\frac{C}{R}}. \tag{3.7}$$

A similar technique is used for sensitivity calculation of supporting transistors and blocking transistors.

Note that $D''(x_0) > 0 \Rightarrow D(x_0)$ is a minimum point. However, this only means that the delay along the current critical path can be reduced by changing the size of this transistor, and does not necessarily mean that the circuit delay can be reduced; the circuit delay is the maximum of all path delays in the circuit, and a change in the size of this transistor could increase the delay along some other path, making a new path critical, as is likely to happen if the transistor size is set to x_0. This is the rationale behind bumping up the size of the transistor, rather than setting it to x_0.

From an optimization viewpoint, the procedure of bumping up the size of the most sensitive transistor could be looked upon in the following way. Let the ith transistor (out of n total transistors) be the one with the maximum sensitivity. Define $\mathbf{e_i} \in \mathbf{R}^n$ as

$$(\mathbf{e_i})_j = \begin{cases} 0 & i \neq j \\ 1 & i = j \end{cases}. \tag{3.8}$$

In each iteration, the TILOS optimization procedure works in the n-dimensional space of the transistor sizes, chooses $\mathbf{e_i}$ as the search direction, and attempts to find the solution to the problem by taking a step of size BUMPSIZE along that direction.

3.3 The Method of Feasible Directions (MFD) Algorithm

3.3.1 Description of the Algorithm

Shyu et al. [SFD88, Shy88] proposed a two-stage optimization approach to solve the transistor size problem. The delay estimation algorithm is identical to that used in TILOS.

In the first stage, the TILOS heuristic described in Section 3.2 is used to generate an initial solution. If the heuristic finds a solution which satisfies the constraints, then the sized-up transistors are used as design parameters. This heuristic approach is not guaranteed to find an optimal solution. However, it can serve as an initial guess solution for a mathematical optimization technique.

In the second stage of the optimization process, the problem is converted into a mathematical optimization problem, and is solved by a Method of Feasible Direction (MFD) [PTM71]. The solution generated by TILOS in the first stage is used as an initial guess for this algorithm. The optimization technique solves a sequence of problems that have a smaller number of design parameters in order to reduce the computational complexity. At first, the transistors on the worst-delay paths (usually more than one) are selected as design parameters. If, with the selected transistors, the optimizer fails to meet the delay constraints, and some new paths become the worst-delay paths, the algorithm augments the design parameters with the transistors on those paths, and restarts the process. However, while this procedure reduces the runtime of the algorithm, one faces the risk of finding a suboptimal solution since only a *subset* of the design parameters is used in each step.

The MFD optimization method proceeds by finding a search direction d, a vector in the n-dimensional space of the design parameters, based on the gradients of the cost function and some of the constraint functions. Once the search direction has been computed, a step along this direction is taken, so that the decrease in the cost and constraint functions is large enough. The computation stops when the length of this step is sufficiently small. A detailed description of the optimization algorithm is given in [PTM71]. This algorithm has the feature that once the feasible region (the set of transistor sizes where all delay constraints are satisfied) is entered, all subsequent improvements will remain feasible.

The algorithm can be summarized in the following pseudo-code:

Use TILOS to size the entire circuit;

While (TRUE) {

    Select $G_1, \cdots, G_k$, the k most critical paths,

    and $X = \{x_i\}$, the set of design parameters

    Solve the optimization problem

$$\begin{aligned} \text{minimize} \quad & \sum_{x_i \in X} x_i \\ \text{such that} \quad & G_i(X) \leq T \; \forall \; i = 1, \cdots, k \\ \text{and} \quad & x_i \geq \text{minsize} \; \forall \; x_i \in X. \end{aligned}$$

    If all constraints are satisfied, exit

}

3.3.2 Practical Implementational Aspects

The Generalized Gradient

For convergence, the MFD requires that the objective and constraint functions be continuously differentiable. However, in digital circuits, where the circuit delay is defined as the maximum of all path delays in the circuit, the delay constraint functions are usually not differentiable. To illustrate that the maximum of two continuously differentiable functions, $g_1(x)$ and $g_2(x)$, need not be differentiable, consider the example in Figure 3.4. The maximum function, shown by the bold lines, is nondifferentiable at x_0.

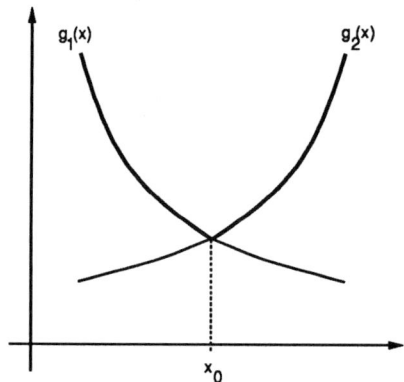

Figure 3.4: Nondifferentiability of the *max* function.

To cope with the nondifferentiability of the constraint functions, a modification of the MFD is used, which employs the concept of the *generalized gradient* [Cla83].

THE MFD ALGORITHM

Before proceeding further, we define the convex hull of m points, $\mathbf{x}_1, \cdots, \mathbf{x}_n$, denoted $co\{\mathbf{x}_1, \cdots, \mathbf{x}_n\}$, in the n-dimensional space, $\mathbf{R}^n$. This is defined as the set of points $\mathbf{y} \in \mathbf{R}^n$ such that

$$\mathbf{y} = \sum_{i=1}^{m} \alpha_i \mathbf{x}_i ; \quad \alpha_i \geq 0 \ \forall \ i, \quad \sum_{i=0}^{m} \alpha_i = 1. \tag{3.9}$$

The convex hull is the smallest convex set that contains the m points. An example of the convex hull of five points in the plane is shown by the shaded region in Figure 3.5.

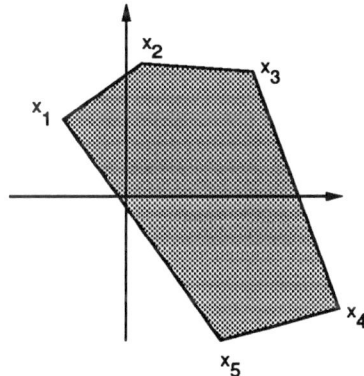

Figure 3.5: The convex hull of five points.

The nondifferentiability of the circuit delay originates with the fact that the maximum function is used to define the circuit delay. Let

$$\psi(\mathbf{x}) = \max_i g_i(\mathbf{x}), \quad i = 1, 2, \cdots, n \tag{3.10}$$

be the circuit delay, where $\mathbf{x}$ is the vector of the design variables (transistor sizes), and the g_i's are the path delays. Let

$$I(\mathbf{x}) = \{i \mid g_i(\mathbf{x}) = \psi(\mathbf{x})\}. \qquad (3.11)$$

If the set $I(\mathbf{x})$ has only one element, then the circuit delay is obviously differentiable at $\mathbf{x}$. However, if $I(x)$ has more than one element, then the circuit delay could be nondifferentiable at $\mathbf{x}$, as illustrated in Figure 3.4. The generalized gradient is introduced to compute descent directions for $\psi(\mathbf{x})$ for these cases. It is formally defined as follows:

$$\partial \psi(\mathbf{x}) = co \{\nabla g_i(\mathbf{x}) \mid i \in I(\mathbf{x})\} \qquad (3.12)$$

where co denotes the convex hull of the vectors $\nabla g_i(\mathbf{x})$, i.e., the set defined by

$$\sum_{i \in I(\mathbf{x})} \mu_i \nabla g_i(\mathbf{x}) = 0 \; ; \quad \sum_{i \in I(\mathbf{x})} \mu_i = 1, \; \mu_i \geq 0, \; i \in I(\mathbf{x}). \qquad (3.13)$$

Note that when $I(\mathbf{x})$ is a singleton set, the generalized gradient reduces to the standard definition of the gradient. If it has more than one element, $\partial \psi(\mathbf{x})$ is a set as shown in Figure 3.6(a).

For effective use in this algorithm, a set which is closely related to the generalized gradient is defined as follows. First, let

$$I_\epsilon(\mathbf{x}) = \{i \mid \psi(\mathbf{x}) - g_i(\mathbf{x}) \leq \epsilon\} \qquad (3.14)$$

THE MFD ALGORITHM

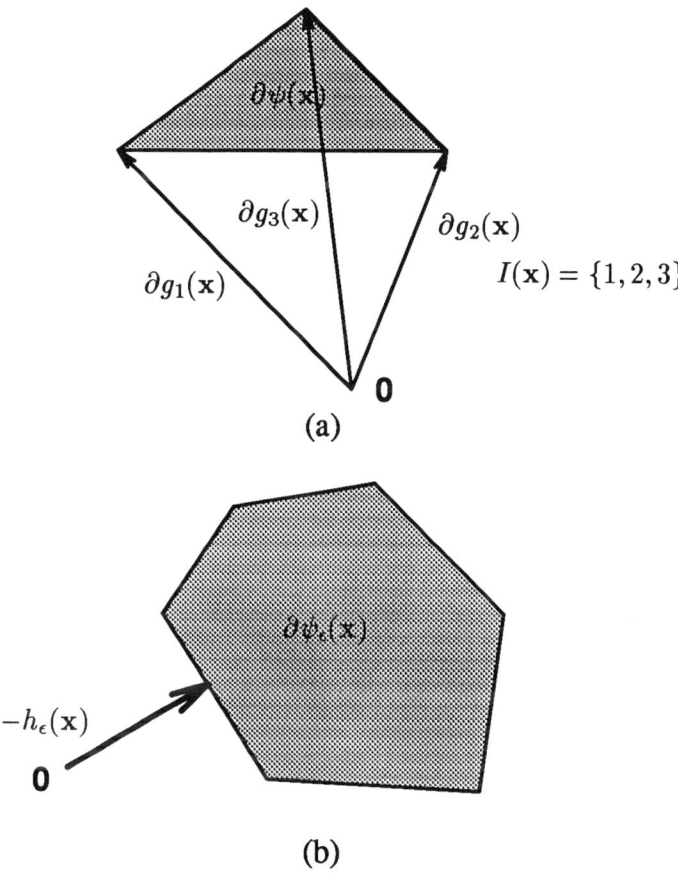

Figure 3.6: Computing the generalized gradient.

be the ϵ-*active* index set, where ϵ is a small positive number. $I_\epsilon(\mathbf{x})$ thus consists of the critical path and the paths whose delays are within ϵ of the critical path delay. Then define the set

$$\partial \psi_\epsilon(\mathbf{x}) = co\left\{\nabla g_j(\mathbf{x}) \mid j \in I_\epsilon(\mathbf{x})\right\}. \tag{3.15}$$

The MFD algorithm here uses the shortest vector in the set

$\partial_\epsilon \psi(\mathbf{x})$ to compute the search direction $h_\epsilon(\mathbf{x})$, i.e.,

$$h_\epsilon(\mathbf{x}) = -Nr(\partial_\epsilon \psi(\mathbf{x})) \qquad (3.16)$$

where Nr denotes the nearest vector to the origin, which is the shortest vector in the set $\partial_\epsilon \psi(\mathbf{x})$. Figure 3.6(b) shows how $h_\epsilon(\mathbf{x})$ is computed graphically.

Scaling

It is important to scale the gradients of the cost and path delay functions. The use of their unscaled values may produce poor descent search directions due to the large difference in magnitude of the gradients of the constraint and objective functions. When a gradient has a magnitude that is much smaller than other gradients, it dominates the search direction. In such a case, the descent direction is unjustly biased away from the other constraints and/or the cost function, since the chosen vector, which is nearest to the origin, will be almost identical to this gradient.

3.4 Lagrangian Multiplier Approaches

As can be seen from the approaches studied so far, the problem of transistor sizing can be formulated as a constrained nonlinear programming problem. Hence, the method of Lagrangian multipliers

3.4. LAGRANGIAN MULTIPLIER APPROACHES

[Lue84] is applicable. A few methods that employ this approach are described in this section.

3.4.1 Early Approaches

Cirit's Method

An approach proposed by Cirit [Cir87] attempts to solve the problem of minimizing the area (defined as the sum of transistor sizes), subject to a given timing specification along a certain path of the circuit. This algorithm was intended to be used as an interactive design aid, whereby the designer would identify the critical path using a timing analyzer, and the optimizer would attempt to meet the delay specification for that path.

The primary disadvantage of this approach is that it ignores interactions between paths. In other words, solving the above problem for a certain path could worsen the delay along some other path. Hence, the solution found by this algorithm is, in general, not optimal.

The AESOP algorithm

AESOP [Hed87] solves two formulations of the transistor sizing problem. The first formulation attempts to minimize the delay, subject to upper- and lower-bound constraints on the transistor sizes. The delay function is expressed as

$$Delay = min\{max D_P(\mathbf{x})\} \quad (3.17)$$

where D_P is the delay of an input-output path, and $\mathbf{x}$ is the vector of transistor sizes. Since Eq. (3.17) is not a continuous function, the following continuous function, first proposed in [RPG77], is minimized instead.

$$min\{smax D_P(\mathbf{x})\} \quad (3.18)$$

where $smax(x_1,\cdots,x_n) = \frac{1}{\lambda}ln(e^{\lambda x_1}+\cdots+e^{\lambda x_n})$. However, as pointed out in [Mar86], this process of 'smoothing' the discontinuous function in Eq. (3.17) corrupts the convergence property of the original problem.

The second formulation minimizes the area, subject to path delay constraints. The user must identify potential critical paths, so that the program takes the constraints corresponding to these paths into consideration. However, for some circuits, the number of such paths could be very large, and in the worst case, could be exponential in the number of components. Since the number of such constraints is large, the runtimes for a Lagrangian-based algorithm could be high.

3.4.2 Marple's Approach

Marple [Mar86] presented a solution to transistor size optimization by using Lagrangian multipliers. This technique uses a new area

3.4. LAGRANGIAN MULTIPLIER APPROACHES

model and employs the idea of introducing intermediate variables to reduce the number of delay constraints from an exponential number to a number that is linear in the circuit size.

Moreover, Marple's approach is based on the work of [Man65], which showed that the point that satisfies the first-order Kuhn-Tucker conditions [Lue84] is a global optimum point for posynomial and weakly convex programs. This avoids the smoothing techniques used in [RPG77, Hed87].

Area Model

Since the representation of circuit area as the sum of transistor sizes is unrealistic, this technique begins with a layout, and performs the optimization for that layout. While such an approach has the disadvantage that it may not result in the minimal area over *all* layouts, it still maintains the feature that the area and delay constraints are posynomials. Hence, the transistor sizing problem (3.1) can be formulated as a convex programming problem, with its accompanying advantages (See Section 2.6), to minimize the area for the given layout schematic. For a given layout, apart from the delay constraints, there also exist some area constraints, modeled by constraint graphs that are commonly used in layout compaction [Boy88]. These constraints maintain the minimum spacing between objects in the layout, which is specified by design rules.

Optimization Technique

The delay of the circuit is modeled by a delay graph, $D(V, A)$ where V is the set of nodes (components) in D, and A is the set of arcs (connections among components) in D. This is the same graph on which the PERT analysis is to be carried out. Let m_i represent the worst-case delay at the output of component i, from the primary inputs. Then for each component, the delay constraint is expressed as

$$m_i + d_j \leq m_j. \tag{3.19}$$

where component $j \in$ fanout(component i), and d_j is the delay of component j.

Thus, the number of delay constraints is reduced from a number that could, in the worst case, be exponential in $|V|$, to one that is linear in $|A|$, by the addition of certain variables. The number of new variables is equal to $|V|$.

These techniques are implemented in COP [MG87], a program for optimization of general CMOS circuits, and in PLATO [MG86], a system for PLA layout generation. The objective is to minimize the area of the layout, subject to area and delay constraints. The area constraints used in COP correspond to the circuit floorplan

3.4. LAGRANGIAN MULTIPLIER APPROACHES

requirements.

Two variations of the Lagrange multiplier approach, the augmented Lagrangian algorithm and the projected Lagrangian algorithm [Lue84, Mar86], are used to solve the problem. The augmented Lagrangian algorithm uses a penalty term that helps to steer the solution towards the feasible region and, hence, has the desirable property of global convergence. The projected Lagrangian method, a quadratic approximation method that is similar to Newton's method, has a fast convergence rate. However, it is not globally convergent, and requires that the initial solution be close enough to the optimum solution in order to converge. COP starts with the augmented Lagrangian technique to take advantage of its global convergence. When the gradient of the Lagrangian becomes less than a certain small number, ϵ, it switches to the projected Lagrangian algorithm, to converge more quickly to the solution.

A subsequent algorithm [Mar89] uses a somewhat different approach. This algorithm interweaves layout compaction with transistor sizing and is implemented in the Tailor design system. As before, the transistor sizing procedure is performed on an actual layout. However, the area constraints here correspond to physical objects within the layout, such as diffusion lines and transistors, rather than the abstract floorplanning blocks used in COP. The complete nonlinear program, including both linear and nonlinear constraints could be solved with a general nonlinear programming method. However, such an approach would not take advantage of any of the structural properties of the problem, and would thus be inefficient and slow. Hence, dedicated algorithms are used for the linear (area) constraints, and nonlinear programming algorithms are employed for the nonlinear (delay) constraints. Tailor's transistor size optimizer finds the initial active linear constraints by using the longest-path algorithm for compaction [MSH88]. Once the active linear constraints are determined, the optimizer proceeds with the nonlinear optimization. Nonlinear optimization is implemented with the augmented Lagrangian algorithm mentioned above. If, at any time during the optimization, it becomes apparent that the wrong linear area constraints are considered as being active, an updating operation is carried out, providing a new and better set of active linear constraints.

3.5 Two-step Optimization

Since the number of variables in the transistor sizing problem, which equals the number of transistors in a combinational segment, is typically too large for most optimization algorithms to handle efficiently, many algorithms choose a simpler route by performing the optimization in two steps. Examples of algorithms that use this idea to solve

3.5. TWO-STEP OPTIMIZATION

the transistor sizing problem are iCOACH [CK91], MOSIZ [DA89], and CATS [HF91].

In the first step in MOSIZ, macromodeling techniques are used to map each component to an equivalent primitive, such as an inverter. The transistor sizing problem on this simplified circuit is then solved. Note that the number of variables is substantially reduced when each component is replaced by a simple primitive, with fewer transistors. The delay of each equivalent inverter, with the transistor sizes obtained above, is taken as the *timing budget* for the component represented by that inverter.

iCOACH uses macromodels for timing analysis of the circuit, and has the capability of handling dynamic circuits. The optimizer employs a heuristic to estimate an *improvement factor* for each gate, which is related to the sensitivity of the gate. The improvement factor depends on the fanin count, fanout count and the worst-case resistive path to the relevant supply rail. The improvement factor is then used to allocate a timing budget to each gate.

In the second step, for each component, a smaller transistor sizing problem is solved, in which the area of the component is minimized, subject to its delay being within its timing budget. The number of variables for each such problem equals the number of transistors within the component, which is typically a small number.

The two steps are repeated iteratively until the solution converges. While this technique has the obvious advantage of reducing the number of design parameters to be optimized, it suffers from the disadvantage that the solution may be nonoptimal. This stems from the simplifications introduced by the timing budget allocation; the timing budget allocated to each component may not be the same as the delay of the component for the optimal solution.

3.6 Other Approaches

Other approaches to solving this problem include

- MOGLO [HNS90], which formulates the problem as a multi-criterion optimization problem, with a convex combination of the power, delay and area as the objective function.

- PROMPT3 [CC89] which uses simulated annealing [KGV83], to choose an optimal set of sized logic cells from a library to solve the transistor sizing problem. In the initial step, a TILOS-like heuristic is used to find an initial solution that satisfies the timing requirements. Next, an area minimization is carried out on a restricted problem, using simulated annealing.

- The approach used by Berkelaar and Jess [BJ90], which uses delay models that are piecewise linear functions of the transistor sizes. Under this assumption, the transistor sizing problem is formulated as a linear programming problem.

- iDEAS [SR90], in which the delay along the critical path is reduced in each iteration. In each iteration, the sizes of transistors on the critical path are considered as variables, and a reduced problem is solved, in which the delay of the critical path is minimized. The minimization uses a cyclic coordinate descent technique, giving a solution vector $\mathbf{x}_{opt} \in \mathbf{R}^m$ for the optimal sizes of the m transistors on the critical path. If $\mathbf{x} \in \mathbf{R}^m$ is the set of transistor sizes in the current circuit, the sizes of the transistors on the critical path are changed by taking a small step in the direction $\mathbf{x}_{opt} - \mathbf{x}$. The iterations continue until the timing specification is satisfied.

3.7 Summary of Previous Approaches

The optimization techniques used by most of the approaches that are described in this chapter are not guaranteed to find the optimal solution to the transistor sizing problem. Most previous transistor sizing algorithms fail to take advantage of the convexity properties of the objective and constraint functions, and use only the unimodality property of the convex program. In this section, we consider the form of the transistor sizing problem given in Eq. (3.1) and point out the limitations of the approaches discussed in Sections 3.2-3.5.

TILOS [FD85] uses essentially a greedy heuristic that attempts to solve the problem by bumping up the size of the most sensitive transistor in each iteration. However, this procedure does not consider interactions between paths while computing the sensitivity. For example, it is possible that an increase in the size of a transistor increases the delay along another (currently noncritical) path, which subsequently becomes critical. In such a case, it may be desirable to reduce the size of that transistor; however, TILOS only permits transistor sizes to be increased.

Shyu's [SFD88] approach uses an initial guess solution provided by TILOS and identifies a reduced parameter space over which the optimization is performed. An optimization is performed over this reduced parameter space, in an attempt to reach the global optimum. However, achieving the global minimum cannot be guaranteed, since all of the variables have not been considered in the optimization.

The simple Lagrangian multiplier approaches, such as Cirit's [Cir87], attempt to minimize the delay along one path at a time, and ignore interactions between paths, and are thereby potentially sub-

3.7. SUMMARY OF PREVIOUS APPROACHES

optimal. The procedure used by AESOP [Hed87] requires the user to identify potential critical paths. The number of such paths is, in the worst case, exponential in the number of components; hence, in the worst case, this problem cannot be solved in a reasonable amount of time. Marple's [Mar86] approach attempted to formulate the delay constraints so as to reduce the number of constraints without making approximations. However, this required the introduction of k additional variables, where k is the number of components in the circuit, and the number of constraints is still equal to the number of lines (connections between gates), which, in the worst case, is of the order $O(k^2)$. While this presents an improvement over having an exponential number of constraints, the number of constraints could still be very large.

The two-step methods of Section 3.5 lose sight of the global picture when they work in a reduced parameter space by considering one variable for each component to obtain an initial solution. Unless the timing budget is correctly allocated in the first step, a sequence of local optimizations based on these assumptions would not lead to the global optimum.

Chapter 4

A Convex Programming Approach to Transistor Sizing

4.1 Introduction

In Chapter 3, the transistor sizing problem was defined, and various approaches that have been used to tackle this problem were described. The chief shortcoming of most of these approaches, as pointed out in Section 3.7, was that the simplifying assumptions made by these algorithms to make the optimization problem more tractable may lead to a suboptimal solution.

As pointed out in Chapter 3, it has widely been recognized that the active area, defined as the sum of transistor sizes, and the delay along a path of the circuit can be represented by *posynomial* functions of the sizes of transistors in the circuit. A posynomial is a function g of a positive variable $\mathbf{x} \in \mathbf{R}^n$ that has the form

$$g(\mathbf{x}) = \sum_j \gamma_j \prod_{i=1}^n x_i^{\alpha_{ij}} \qquad (4.1)$$

where the exponents $\alpha_{ij} \in \mathbf{R}$ and the coefficients $\gamma_j > 0$. Such a function has the useful property that it can be mapped onto a convex function through an elementary variable transformation, $(x_i) = (e^{z_i})$ [Eck80].

For our purposes, the delay of a circuit is defined to be the maximum of the delays of all paths in the circuit. Hence, it can be formulated as the maximum of posynomial functions. This is mapped by the above transformation onto a maximum of convex functions, which is also a convex function. The area function is also a posynomial, and is transformed into a convex function by the same mapping. Therefore, the optimization problem defined in (3.1) is mapped to a *convex programming* problem, i.e., a problem of minimizing a convex function over a convex constraint set. Due to the unimodal property of convex functions over convex sets, any local minimum of (3.1) is also a global minimum.

In this chapter, we show how the transistor sizing problem as defined in (3.1) is tackled. The other formulations mentioned in Section 3.1 can also be handled using the same approach.

4.1. INTRODUCTION

The delay estimation algorithm used in this approach has been described in Section 2.7 An efficient convex programming method [Vai89] is used for global optimization over the parameter space of all transistor sizes in a combinational subcircuit, thereby solving the problem *exactly*. The algorithm works in the space $\mathbf{R}^n$, where n is the number of transistors in the circuit. Each point in the space corresponds to a particular set of transistor sizes. The objective of the algorithm is to identify the point which corresponds to the globally optimum set of transistor sizes.

The algorithm starts by bounding the convex domain by an initial polytope. By using a special cutting plane technique, the volume of this polytope is shrunk in each iteration, while ensuring that the optimal solution lies within the boundary of the reduced polytope. The iterative procedure stops when the volume of the polytope becomes sufficiently small. A more complete description is given in Section 4.2. Experimental results to illustrate the efficacy of this technique are presented in Section 4.3.

4.2 The Convex Programming Algorithm

The objective of the algorithm is to solve the following transistor sizing problem

$$\text{minimize} \quad Area(\mathbf{x}) = \sum_{i=1}^{n} x_i$$

$$\text{subject to} \quad Delay(\mathbf{x}) \leq T_{spec} \quad (4.2)$$

where the delay $Delay(\mathbf{x})$ is maximum of delays along all paths to a primary output node of the circuit. By making the variable transformation

$$(x_i) = (e^{z_i})$$

the original transistor sizing problem (4.2) of minimizing a posynomial area function over posynomial constraints becomes

$$\text{minimize} \quad Area(\mathbf{z}) = \sum_{i=1}^{n} e^{z_i} \quad (4.3)$$

$$\text{subject to} \quad D(\mathbf{z}) \leq T_{spec}.$$

Note that under this transformation, the delay along a path has the form

$$\sum_j \gamma_j e^{\sum_{i=1}^{n} \alpha_{ij} z_i}$$

which is a convex function. Since the circuit delay is defined to be the maximum of all path delays, and the maximum of convex functions

4.2. THE CONVEX PROGRAMMING ALGORITHM

is also convex, $D(\mathbf{z})$ is a convex function. Moreover, $Area(\mathbf{z})$ is also known to be a convex function of $\mathbf{z}$. Hence, (4.3) is a problem of minimizing a convex function over a convex set of constraints, i.e., a *convex programming* problem. The algorithm proposed by Vaidya in [Vai89] provides an efficient technique for solving a convex programming problem such as Eq. (4.3).

We first define the *feasible set*

$$S = \{\mathbf{z} \in \mathbf{R}^n : D(\mathbf{z}) \leq T_{spec}\} \quad (4.4)$$

i.e., the set of points that satisfies the delay constraints. Let $\mathbf{z}_{opt}$ be the solution to (4.3). $\mathbf{z}_{opt}$ is then the point within S that minimizes the area function.

Initially, a polytope (i.e., a bounded polyhedron) P in $\mathbf{R}^n$ that contains $\mathbf{z}_{opt}$ is chosen. The objective of the algorithm is to start with a large polytope, and in each iteration, to shrink its volume while keeping $\mathbf{z}_{opt}$ within the polytope, until the polytope becomes sufficiently small. The polytope is represented in the form

$$P = \{\mathbf{z} : A\mathbf{z} \geq \mathbf{b}\} \quad (4.5)$$

where $A \in \mathbf{R}^{m \times n}$ and $\mathbf{b} \in \mathbf{R}^m$. Here, m denotes the number of linear inequality constraints describing the polytope. The initial polytope P may, for example, be selected to be an n-dimensional box described by the set

$$\{\mathbf{z} : \log_e(x_{min}) \leq z_i \leq \log_e(x_{max})\} \qquad (4.6)$$

where x_{min} and x_{max} are the user-specified minimum and maximum allowable transistor sizes, respectively.

The algorithm proceeds iteratively as follows.

Step 1

Find a *center* $\mathbf{z}_c$ deep in the interior of the current polytope P by using a technique that will be described later.

Step 2

Invoke an *oracle* to determine whether or not the center $\mathbf{z}_c$ lies within the feasible region S.

From the definition of S, the oracle is simply a routine that invokes the delay estimator described in Section 2.7, with the transistor sizes $x_i = e^{(\mathbf{z}_c)_i}$, to determine whether or not the delay requirement is met.

If the point $\mathbf{z}_c$ lies outside S, it is possible to find a *separating hyperplane* passing through $\mathbf{z}_c$ that divides the polytope P into two parts, such that S lies entirely in the part satisfying the constraint

$$\mathbf{c}^T \mathbf{z} \geq \beta \qquad (4.7)$$

4.2. THE CONVEX PROGRAMMING ALGORITHM

where

$$\mathbf{c} = -[\nabla D_{critpath}(\mathbf{z})]^T \qquad (4.8)$$

is the negative of the gradient of the critical path delay (constraint) function, and

$$\beta = \mathbf{c}^T \mathbf{z}_c. \qquad (4.9)$$

The separating hyperplane described above corresponds to the tangent plane to the path delay along the critical path. Note that the discontinuity of the derivative of the circuit delay function does not affect matters, since we only need with the gradient of a path delay, a continuous function.

If the point $\mathbf{z}_c$ lies within the feasible region S, then there exists a hyperplane that divides the polytope into two parts such that $\mathbf{z}_{opt}$ is contained in one of them satisfying the constraint (4.7) with

$$\mathbf{c} = -[\nabla Area(\mathbf{z})]^T \qquad (4.10)$$

being the negative of the gradient of the area (objective) function, and β is once again defined by (4.9).

Step 3

In either case, add the constraint (4.7) to the current polytope to give a new polytope that has roughly half the original volume.

Step 4

Repeat the process until the polytope is sufficiently small.

Example

We illustrate the algorithm by using it to solve the following problem

$$\text{minimize} \quad f(x_1, x_2)$$

$$\text{such that} \quad (x_1, x_2) \in S \tag{4.11}$$

where S is a convex set and f is a convex function.

The shaded region in Figure 4.1(a) corresponds to S, and the dotted lines show the level curves of the function f. The point $\mathbf{x}^*$ is the solution to this problem. The procedure is as follows:

- The expected solution region is bounded by a closed initial polytope, which is a rectangle in two dimensions. This is shown in Figure 4.1(a).
- The center, $\mathbf{z}_c$, of this rectangle is found.
- The oracle is invoked to determine whether or not $\mathbf{z}_c$ lies within the feasible region. In this case, it can be seen that $\mathbf{z}_c$ lies outside the feasible region. Hence, the gradient of the constraint function is used to construct a hyperplane through $\mathbf{z}_c$, such that the polytope is divided into two parts of roughly equal

4.2. THE CONVEX PROGRAMMING ALGORITHM

volume, one of which contains the solution $\mathbf{x}^*$. This is illustrated in Figure 4.1(b), where the shaded region corresponds to the updated polytope.

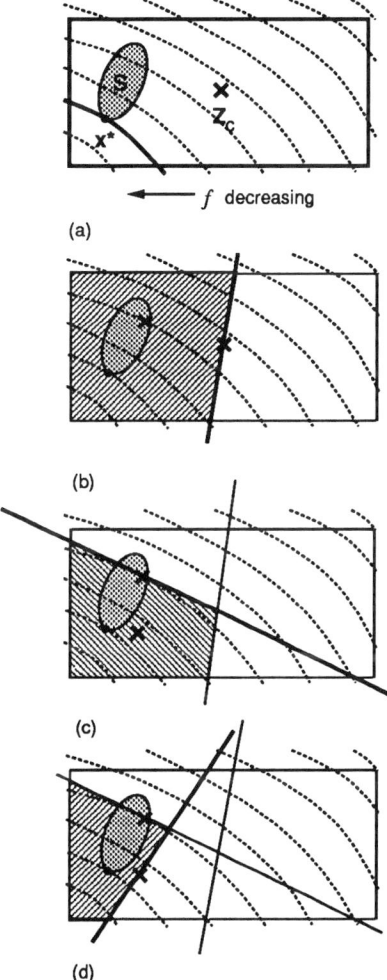

Figure 4.1: Example illustrating the convex programming algorithm.

- The process is repeated on the new smaller polytope. Its center lies inside the feasible region; hence, the gradient of the objective function is used to generate a hyperplane that further shrinks the size of the polytope, as shown in Figure 4.1(c).

- The result of another iteration is illustrated in Figure 4.1(d).
- The process continues until the polytope has been shrunk sufficiently according to a user-specified criterion.

It can be seen that the key parts of this algorithm are

(1) finding the center $\mathbf{z}_c$ of the existing polytope P,

(2) invoking the oracle and generating gradient functions in (4.8) and (4.10) above, and

(3) deciding when to terminate the algorithm.

4.2.1 Finding the Center of the Polytope

We would like to find a point inside a polytope that satisfies the property that any separating hyperplane drawn through it divides the original polytope into two parts of approximately equal volume. Since finding such a point is difficult [Vai89], we settle for finding a point that is reasonably deep within the interior of the polytope, and can be found through relatively inexpensive computation.

Consider a polytope P defined by (4.5), and let $\mathbf{a}_i^T$ be the ith row of the $m \times n$ matrix A, and b_i be the ith element of the m-dimensional

4.2. THE CONVEX PROGRAMMING ALGORITHM

vector b. The center z_c, is taken to be the vector that minimizes the following *log-barrier function*

$$F(\mathbf{z}) = -\sum_{i=1}^{m} \log_e(\mathbf{a_i}^T\mathbf{z} - b_i). \qquad (4.12)$$

Note that near the boundary of the polytope, $F(\mathbf{z}) \to \infty$ and its value decreases as one moves *deeper* into the interior of the polytope. The value of $F(\mathbf{z})$ is undefined outside the boundary of the polytope. F is a convex function of $\mathbf{z} \in P$, with a $1 \times n$ gradient vector

$$\nabla F(\mathbf{z}) = -\sum_{i=1}^{m} \frac{\mathbf{a_i}^T}{(\mathbf{a_i}^T\mathbf{z} - b_i)} \qquad (4.13)$$

and an $n \times n$ Hessian matrix

$$H(\mathbf{z}) = \nabla^2 F(\mathbf{z}) = \sum_{i=1}^{m} \frac{\mathbf{a_i}\mathbf{a_i}^T}{(\mathbf{a_i}^T\mathbf{z} - b_i)^2}. \qquad (4.14)$$

Since the initial polytope is a box, its center is easy to find. In each subsequent iteration, a constraint of the form $\mathbf{c}^T\mathbf{z} \geq \beta$ is added to the previous polytope whose center is found iteratively using Newton's method [Lue84] as follows. The initial point $\mathbf{z}_0$ for Newton's method is found by moving halfway to the closest boundary in the direction $\mathbf{c}$. The point $\mathbf{z}_0$ thus obtained is guaranteed to be in the interior of the new polytope.

The Newton's method for finding the center z_c then generates iterates of the form

$$z_{k+1} = z_k + t^* \xi_k \qquad (4.15)$$

for $k = 0, 1, 2, \cdots$, until convergence, where ξ_k is the *Newton direction* at z_k given by

$$\xi_k = -H^{-1}(z_k)[\nabla F(z_k)]^T = -[\nabla^2 F(z_k)]^{-1}[\nabla F(z_k)]^T \qquad (4.16)$$

and t^* is the point that minimizes the one-dimensional function

$$\phi(t) = F(z_k + t\xi_k) \qquad (4.17)$$

and is obtained by performing a one-dimensional line search.

Note that the process of computing a Newton direction by (4.16) involves the inversion of an $n \times n$ Hessian matrix which takes $O(n^3)$ time and can prove to be rather expensive. This computation can be cut down by maintaining the inverse of an approximate Hessian $\hat{H}$ via rank-one updates [Lue84] as described below, and using an approximate Newton direction $\hat{\xi}_k$ instead of ξ_k in the line search. Note that using an approximate Newton direction instead of the exact one essentially does not affect the convergence properties of the center-finding algorithm [Vai89].

4.2. THE CONVEX PROGRAMMING ALGORITHM

4.2.2 Rank-one Updates

Let z_k be the point at the beginning of the $(k+1)^{th}$ iteration of Newton's method for finding the center z_c of the polytope P described by (4.5).

Two methods for maintaining the approximate Hessian, using rank-one updates [Lue84], are outlined below.

Method 1

The Hessian at z_k may be written as

$$H(z_k) = \nabla^2 F(z_k) = \sum_{i=1}^{m} \frac{\mathbf{a_i}\mathbf{a_i}^T}{(\mathbf{a_i}^T z_k - b_i)^2} = A^T \Lambda A \qquad (4.18)$$

where $\Lambda \in \mathbf{R}^{m \times m}$ is a diagonal matrix, with $(\Lambda)_{ii} = (\mathbf{a_i}^T z_k - b_i)^{-2}$.

Let $\delta_1, \delta_2 > 0$ be small parameters. An approximate Hessian is given by

$$\hat{H} = A^T \hat{\Lambda} A \qquad (4.19)$$

where $\hat{\Lambda} \in \mathbf{R}^{m \times m}$ is a diagonal matrix such that at the start of the $(k+1)^{th}$ iteration, the ith diagonal entry of $\hat{\Lambda}$, $\hat{\lambda}_{ii}$ satisfies the condition

$$(\mathbf{a_i}^T z_{k-1} - b_i)^{-2} \delta_1^{-1} \leq \hat{\lambda}_{ii} \leq (\mathbf{a_i}^T z_{k-1} - b_i)^{-2} \delta_2, \; \forall \; 1 \leq i \leq m.$$

$$(4.20)$$

We maintain an approximate inverse Hessian, $\mathcal{H}^{-1}$. The following rank-one correction procedure is used to update $\mathcal{H}^{-1}$ at the beginning of the $(k+1)^{th}$ iteration.

```
For each i = 1 , 2 , ··· , m {
    if (λ_ii < (a_i^T z_k − b_i)^−2 δ_1^−1) or
       (λ_ii > (a_i^T z_k − b_i)^−2 δ_2) then {
        ω = (a_i^T z_k − b_i)^−2 − λ_ii
        λ_ii = (a_i^T z_k − b_i)^−2
        e = H^−1 a_i
        μ = ω(1 + ω a_i^T e)^−1
        H^−1 = H^−1 − μ e e^T
    }
}
```

One of two procedures may be used to calculate the approximate Newton direction:

<u>Scheme (a)</u>: Maintaining a more accurate $\mathcal{H}^{-1}$, and setting

$$\hat{\xi}_k = -\mathcal{H}^{-1}(\nabla F(z_k))^T. \qquad (4.21)$$

It can easily be verified that each rank-one update to $\mathcal{H}^{-1}$ is of complexity $O(n^2)$. Typically, the number of updates to $\hat{\Lambda}$ per iteration is less than $O(\sqrt{n})$, which reduces the average cost of an iteration of the center-finding algorithm from $O(n^3)$ to $O(n^{2.5})$.

<u>Scheme (b)</u>: Maintaining a more approximate $\mathcal{H}^{-1}$, and using it as a preconditioner for a preconditioned conjugate gradient method [GL89] that solves

$$H\hat{\xi}_k = -(\nabla F(z_k))^T. \qquad (4.22)$$

4.2. THE CONVEX PROGRAMMING ALGORITHM

This method trades off the cost of maintaining $\mathcal{H}^{-1}$ accurately against the cost of performing a few iterations of the preconditioned conjugate gradient method.

For Scheme (a) for maintaining an approximate inverse Hessian described above, the parameter δ_1 above is typically chosen to be around 1.5, while δ_2 is set to about 5. For Scheme (b), typical values for δ_1 and δ_2 are 3 and 20, respectively.

The reason why δ_2 is set to be larger than δ_1 is as follows. When ω is positive (i.e., when δ_2 determines whether an update is to be made or not), the denominator of μ, equal to $(1+\omega \mathbf{a_i}^T \mathbf{e})$, is relatively large; hence, numerical errors in the calculation of μ are damped out. In the case in which ω is negative (i.e. the update decision is dependent on the value of δ_1), the denominator of μ could grow smaller as δ_1 increases, and a large δ_1 could lead to an amplification of numerical errors. Therefore, the choice of δ_2 may be more liberal than that of δ_1.

In each of these two methods, it suffices to maintain $\mathcal{H}^{-1}$; it is not even necessary to explicitly find $\hat{H}$.

Method 2

The Hessian at $\mathbf{z}_k$ may also be written as

$$H = \Gamma + U^T \Omega U \qquad (4.23)$$

Let p be the number of additional planes added to the initial polytope, the box, described by (4.6). $\Gamma \in \mathbf{R}^{n \times n}$ is the Hessian at z_k due to the planes of this box only, and is a diagonal matrix. The ith diagonal element of Γ, denoted γ_{ii}, is given by

$$\gamma_{ii} = \left[[z_{k,i} - x_{min}]^{-2} + [x_{max} - z_{k,i}]^{-2} \right]^{-1}.$$

The rows of $U^T \in \mathbf{R}^{p \times n}$ correspond to the planes added to the initial polytope, i.e., the $(2n+1)^{th}$ to the mth rows of A^T. $\Omega \in \mathbf{R}^{p \times p}$ is a diagonal matrix, whose diagonal entries correspond to the last p diagonal entries of the matrix Λ in (4.18).

We may now write

$$H^{-1} = \Gamma^{-1} - \Gamma^{-1} U \left[\Omega^{-1} + U^T \Gamma^{-1} U \right]^{-1} U^T \Gamma^{-1} \quad (4.24)$$

$$= \Gamma^{-1} - \Gamma^{-1} U C^{-1} U^T \Gamma^{-1} \quad (4.25)$$

where $C = \Omega^{-1} + U^T \Gamma^{-1} U$

We maintain an approximation, $\mathcal{C}^{-1}$, to C^{-1}. An approximate inverse Hessian is then given by

$$\mathcal{H}^{-1} = \Gamma^{-1} - \Gamma^{-1} U \mathcal{C}^{-1} U^T \Gamma^{-1}. \quad (4.26)$$

As in Method 1, it suffices to maintain the approximate inverse of C; it is not necessary to explicitly store C itself. The approximate Hessian or the approximate inverse Hessian are never explicitly maintained. The search direction is found by computing $\xi =$

4.2. THE CONVEX PROGRAMMING ALGORITHM

$-\mathcal{H}^{-1}[\nabla F(\mathbf{z}_k)]^T$, which involves multiplication of the expression (4.26) for $\mathcal{H}^{-1}$ by a $n \times 1$ vector. The cost of this computation can be seen to be $O(np)$, if $n \gg p$, i.e., the number of added planes is much less than the problem dimension. This is the case for large problems; hence, the use of this method would speed up the computation substantially for large problems.

If the number of additional planes, p has not changed since the last calculation of $\mathcal{C}^{-1}$, all that is required to obtain the new $\mathcal{C}^{-1}$ is a set of rank-one updates. If a new plane is added, a method outlined in [Vai90] may be used to update $\mathcal{C}^{-1}$. The method involves a rank-one update and a few additional operations to incorporate the effect of the newly added plane. As before, one of two schemes may be used to calculate the approximate Newton direction:

Scheme (a): Maintaining a more accurate $\mathcal{C}^{-1}$, and setting

$$\hat{\xi}_k = -\mathcal{H}^{-1}(\nabla F(\mathbf{z}_k))^T. \qquad (4.27)$$

Scheme (b): Maintaining a more approximate $\mathcal{C}^{-1}$, and using it as a preconditioner for a preconditioned conjugate gradient iteration that solves

$$H\hat{\xi}_k = -(\nabla F(z_k))^T. \qquad (4.28)$$

It may be noted that the preconditioned conjugate gradient does not require an approximate H or $\mathcal{H}^{-1}$ explicitly, but multiplies $\mathcal{H}^{-1}$

by an $n \times 1$ vector. This operation is computationally cheap when p is small.

It has been found experimentally that Scheme (b) of Method 2 gives the best overall results.

4.2.3 One-dimensional Line Search

Once the Newton direction ξ_k of (4.16) has been found, the value of t^* that minimizes the one-dimensional function $\phi(t)$, defined by (4.17), is obtained as follows. First, the allowable values of t are bounded by t_{min} and t_{max}, where t_{max} is found by computing the distance from the point z_k to the nearest boundary of the polytope along the ξ_k direction. The derivative of ϕ in the interval $[0, t_{max}]$ can be shown to be

$$\phi'(t) = \nabla F(z_k + t\xi_k) \cdot \xi_k = -\sum_{i=1}^{m} \frac{s_i}{s_i t + r_i} \qquad (4.29)$$

where $s_i = \mathbf{a_i}^T \xi_k$ and $r_i = \mathbf{a_i}^T z_k - b_i$ for each $i = 1, 2, \cdots, m$. Note that

$$\phi'(0) = \nabla F(z_k) \cdot \xi_k = -\nabla F(z_k) \cdot H^{-1} \cdot [\nabla F(z_k)]^T < 0 \qquad (4.30)$$

since the Hessian of F, a convex function is positive definite. Also,

$$\lim_{t \to t_{max}} \phi'(t) > 0. \qquad (4.31)$$

4.2. THE CONVEX PROGRAMMING ALGORITHM

As a result of (4.30) and (4.31), and since the function ϕ is convex in the interval $[t_{min}, t_{max}]$, t_{min} can be set to 0, and a simple bisection search can be used to find t^* at which $\phi'(t^*) = 0$ as follows :

```
repeat {
    t* = (t_min + t_max)/2
    if ( φ'(t*) and φ'(t_min) are of opposite sign )
        t_max = t*
    else
        t_min = t*
}
until ( | φ'(t*) | < ε )
```

where ϵ is a small positive number.

4.2.4 Generation of Hyperplanes

When the center $\mathbf{z}_c$ of a polytope lies within the feasible region S, the gradient of the area function is required to generate the new hyperplane passing through the center. The area function of (4.3) has a gradient at the point $\mathbf{z}$ given by

$$\nabla Area(\mathbf{z}) = [e^{z_1}, e^{z_2}, \cdots, e^{z_n}]. \qquad (4.32)$$

In the case when the center $\mathbf{z}_c$ lies outside the feasible region S, the gradient of the critical path delay function $D_{critpath}(\mathbf{z}_c)$, given by Eq. (4.1), is required to generate the new hyperplane that is to be added. For each $k = 1, 2, ..., n$, the kth component of the required gradient vector at a point z is given by (see Eq. (4.1))

$$[\nabla D_{critpath}(\mathbf{z})]_k = \sum_j \gamma_j \alpha_{kj} e^{\sum_{i=1}^n \alpha_{ij} z_i}. \tag{4.33}$$

Note that the transistors in the circuit can contribute to the kth component of the gradient of the delay function in either of two ways:

(i) If the kth transistor is a critical, supporting or blocking transistor (as defined in Section 3.2), or

(ii) If the kth transistor is a capacitive load for some critical transistor.

Transistors that satisfy neither of these two requirements have no contribution to the gradient of the delay function.

4.2.5 Termination Criterion

The algorithm should be terminated when the volume of the final polytope is sufficiently small. In practice, near the optimum, the polytope becomes flat in the direction normal to the gradient of the area. A practical termination criterion uses this property.

From the current center, z_c, let z_1 and z_2 be the two nearest points on the boundary of the polytope, in the direction of the positive and negative gradients of the area, respectively. The difference between the areas of the circuit corresponding to the transistor sizes at z_1 and z_2 provides a measure of the flatness of the polytope in

4.3. EXPERIMENTAL RESULTS

the direction of the area gradient. Hence, the termination criterion is taken to be

$$\frac{|Area(\mathbf{z}_1) - Area(\mathbf{z}_2)|}{Area(\mathbf{z}_c)} < \epsilon \qquad (4.34)$$

where ϵ is a small user-specified number (a reasonable default value is 0.01).

4.3 Experimental Results

The algorithms described in the previous sections have been implemented in iCONTRAST (**i**llinois **C**onvex **O**ptimization-based **N**ovel **TRA**nsistor **S**izing **T**ool). The program, written in C, now consists of approximately 8000 lines of code.

The input to the program is a SPICE deck that gives a transistor-level netlist of the circuit. In the preprocessing stage, the circuit is first divided into channel-connected components. Next, latches in the circuit are identified. The circuit is divided into combinational subcircuits that lie between latches, and the delay constraints for each such subcircuit are determined. The main body of the procedure carries out a convex optimization on each combinational subcircuit.

It must be mentioned here that for our experimental results, the approximate Hessian for finding the Newton direction was maintained using Scheme (b) of Method 2 described in Section 4.2.2.

Table 4.1 : Circuits used to evaluate iCONTRAST

Circuit	Description	Trans. Count	Unsized Area (μm)	Unsized Delay
Inv6	6-inverter Chain	12	21.6	7.0ns
Inv10	10-inverter Chain	20	36.0	12.6ns
Tree	Tree of NAND gates	28	50.4	4.0ns
Add2	2-bit Adder	52	93.6	24.2ns
Add8	8-bit Adder	208	374.4	109.8ns
Add32	32-bit Adder	832	1497.6	452.5ns
Seq	Sequential circuit	244	439.2	26.9ns

4.3. EXPERIMENTAL RESULTS

A set of test circuits described in Table 4.1 was used to evaluate the performance of iCONTRAST. The entries under *unsized area* and *unsized delay* correspond to the area and delay when all transistors in the circuit are set to the minimum size. In the case of the sequential circuit, the delay refers to the maximum delay allowable for a combinational stage of the sequential circuit. It may be noted that the word "area" refers to the sum of transistor sizes. The technology parameters used here correspond to a submicron technology.

Table 4.2 shows the area of the circuit after it has been sized by iCONTRAST to meet a delay specification, T_{spec}, and the execution time on a Sun Sparcstation I. Since our method solves the underlying convex programming problem exactly, the areas shown here correspond to the globally optimum solution to the transistor sizing problem, with an accuracy that is dictated by the tightness of the user-specified termination criterion.

Consider, for example, the results on the example circuit, Add8. As seen in Table 4.1, the unsized area and delay for this circuit are 374.4 μ m, and 109.8 ns respectively. The area penalty required to achieve a relatively loose delay specification such as the first one, 100 ns, is not very large; the active area of the sized circuit is only 11% larger than the unsized circuit. As the delay specification becomes tighter, the area penalty increases non-linearly; to achieve a delay specification of 40 ns, the active area of the sized circuit is 182 % larger than that of the unsized circuit. A similar trend is visible for each of the other example circuits in Table 4.2.

In a comparison with TILOS [FD85, DFH89, Fis92], when the delay specification was loose, the area of the TILOS-sized circuit was close (within a few tenths of a percent) to the optimal one obtained using the iCONTRAST algorithm. However, as the delay specification was made tighter, the TILOS solution moved away from the optimal one.

Figure 4.2 shows the variation of transistor sizes in a seven-stage inverter chain. The minimum transistor size allowed here is 1.8 μm. For relatively loose delay specifications, it can be seen that the stages closer to the output of the chain are made considerably large, while those towards the input remain relatively unaffected. As T_{spec} (given in ns) is made tighter, the transistor sizes tend to increase more drastically from the input stage to the output stage. The sizes of transistors in the input stage are restricted by the contribution of the resistance of the source that drives the first stage. The variation of sizes in the p-transistors in each stage stages is illustrated in Figure 4.2(a); the variation of n-transistor sizes follows the same trend, as shown in Figure 4.2(b).

A caveat is in order here: one should curb an instinctive tendency

4.3. EXPERIMENTAL RESULTS 137

Table 4.2 : Results on sizing various circuits using iCONTRAST

Circuit	T_{spec}	Sized Area (μm)	Execution Time (Sun Sparcstation I)
Inv6	5.0ns	29.2	1.2s
	4.0ns	40.9	1.7s
	3.0ns	72.2	2.0s
	2.0ns	244.8	2.9s
Inv10	10.0ns	45.2	2.5s
	8.0ns	62.3	2.4s
	6.0ns	110.0	4.1s
	5.0ns	177.4	5.2s
	4.5ns	251.0	6.6s
Tree	3.5ns	58.8	11.8s
	3.0ns	74.8	12.3s
	2.5ns	104.5	14.1s
	2.0ns	174.7	18.3s
	1.5ns	407.0	20.6s
Add2	18.0ns	114.3	33.5s
	15.0ns	132.0	34.0s
	12.0ns	167.3	45.9s
	10.0ns	198.6	60.7s
	8.0ns	247.1	101.6s
	7.0ns	459.6	160.3s
Add8	100.0ns	414.6	18.2m
	80.0ns	491.1	12.3m
	60.0ns	692.9	11.4m
	40.0ns	1430.3	41.7m
Add32	350.0ns	1909.5	420.9m
	250.0ns	2866.5	456.9m
	200.0ns	4329.6	543.5m
Seq	20.0ns	498.5	169.6s
	15.0ns	633.9	258.7s
	10.0ns	1125.8	429.4s

to compare these variations with the smooth exponential variations of Mead and Conway [MC80], since the two problems are not the same. The Mead-Conway problem principally differs from ours in the following respects:

(a) The objective of their problem is to minimize the number of stages and the circuit delay. In our problem, the circuit topology, and hence, the number of stages is fixed.

(b) The Mead-Conway approach uses a simpler delay model.

4.4 Summary

We have presented a convex programming approach to solving the transistor sizing problem. This approach is guaranteed to find the global minimum solution to the problem. A major advantage is that the delay constraints are incorporated implicitly, and do not have to be explicitly enumerated. Ensuring that the delay of the circuit satisfies the specification is equivalent to ensuring that the delay along each path of the circuit satisfies the specification. Since the number of paths in the circuit could be exponentially large, the number of constraints could grow exponentially with the circuit complexity. Although techniques for reducing the number of constraints exist [Mar89], they involve the introduction of additional variables. Also, the number of constraints could still be quadratic in the worst

4.4. SUMMARY

case. An advantage of the convex programming technique is that no additional variables have to be introduced. The complexity of the algorithm is dependent on the number of variables, the size of the initial polytope, and the termination criterion, and is independent of the number of convex constraints. Moreover, the discontinuities in the circuit delay function do not require special treatment from the algorithm, as in [SFD88]. The algorithm, implemented as a C program, has been tested on purely combinational circuits with up to 832 transistors, and on a sequential circuit. The convex programming approach outperforms other heuristic approaches especially when the timing specifications become tight.

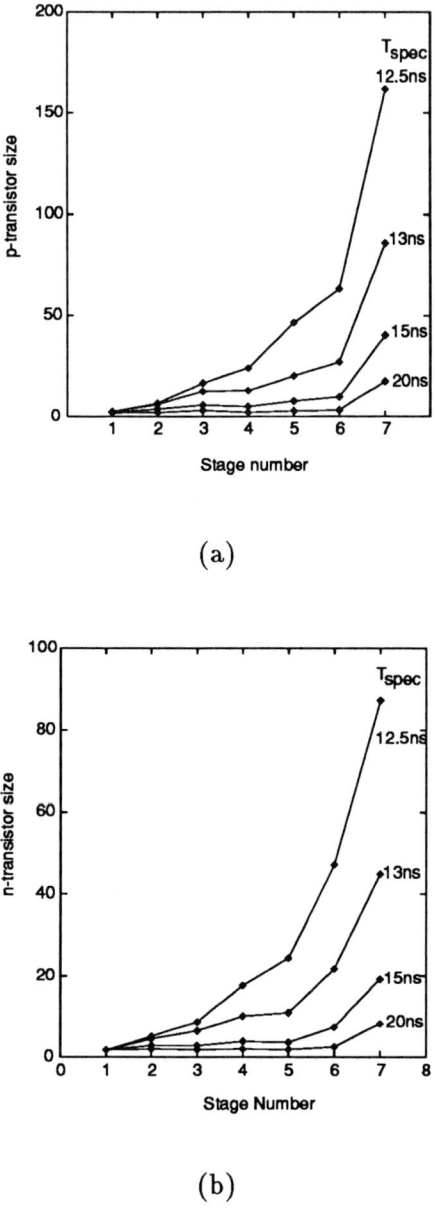

Figure 4.2: (a) iCONTRAST's timing analyzer vs. SPICE (b) An Elmore delay based timing analyzer without iCONTRAST's enhancements vs. SPICE

Chapter 5

Global Routing Using Zero-one Integer Linear Programming

5.1 Introduction

In recent years, the Application Specific Integrated Circuits (ASIC) market has been dominated by gate arrays and their improved form – the sea-of-gates (SOG) arrays [DD89]. The latter provides the advantages of quick turnaround times, high packing density and high performance circuits. With the introduction of large, *channelless* SOG arrays, conventional routers may no longer be able to handle the ever-increasing complexity of the VLSI interconnection problem. In this chapter, a new global router using Zero-one Integer Linear Programming (0-1 ILP) is presented for the case of sea-of-gates arrays. This global router can be used efficiently in high-performance custom layouts, as will be shown in Sections 5.5 and 6.4.

It has long been recognized that one of the most troublesome problem in VLSI routing is the *ordering* of the nets to be wired [MS84, Sou79]. In particular, routing for channelless sea-of-gate arrays demands that all of the wiring be done *over-the-cell*, and therefore suffers from this net-ordering dilemma. The global routing technique [Lek90] using 0-1 ILPs that is proposed in this chapter overcomes this dilemma, and takes into account the constraints imposed by *all* of the nets and the routing environment *simultaneously*.

Integer Linear Programming (ILP) is an established technique, and some researchers have applied it to VLSI routing problems [HK85, HS85a, NRT87, BP83]. With a proper mapping of the global routing problem into a 0-1 linear integer programming (0-1 ILP), the difficulties in combinatorial problems, conflicting requirements, and constraints can be resolved efficiently. It amounts to maximizing the total benefit of selecting a particular set, from many possible routing alternatives, while meeting all routing constraints. Since for each net, only one routing plan can be chosen, binary variables can be assigned to candidate plans with the constraint that the sum of the binary variables for each net should be one. The benefit function can be expressed either in terms of the total wire length or,

5.1. INTRODUCTION

under performance-driven layout, in terms of the weighted sum of wire lengths. In [BP83], net routing was achieved by using 2-by-2 grid cells. Nets were first classified into types and net embeddings into patterns. LP solutions were used to determine how many nets of each net classification should be used for each pattern, but the precise assignment of patterns to nets was not carried out. Also, as pointed out in [HK85], VLSI global routing problems were often too large to be solved directly by ILP methods, and thus resorted heavily on course heuristics. Problems in net decomposition, due to track sharing and the occurrence of cycles in the resulting routing [NRT87], also existed.

Unlike previous work on channelless sea-of-gates global routing, this router addresses real SOG array structures and takes into account the unique features of SOG structures [DD89, PT89, Lau87]. Applying this approach to other SOG array structures would be a straightforward task. In addition, this routing algorithm can handle global routing problems in large SOG arrays, which presents a significant improvement over the approach which performed the global routing *within* SOG modules of up to several hundred transistors [DDK91]. Other features of the ILP-based global router have been motivated by the following observations:

1. *The construction of Rectilinear Steiner Trees (RSTs) [Han66]*

is NP-complete:

Although a heuristic linear time algorithm to construct RSTs has been reported [CRS90], the Steiner tree found is restricted to lie within the rectangle bounded by the terminals. This loss in flexibility may hinder the ability to create routings that avoid obstacles in channelless SOG arrays. The proposed router will avoid the RST construction step completely. Instead of forming the entire RST at once, a few path *segments* of each net can be routed concurrently. This process is repeated until all of the pins in each net are completely connected to each other. This can be achieved by decomposing the global routing process into different *phases* to handle routing of different kinds of path segments—vertical, horizontal and unrestricted.

2. *It is advantageous to implement a Meet-in-the-Middle approach:*
   Conventional top-down and bottom-up hierarchical approaches are merged into a hybrid approach. The intrinsic shortcoming of a purely top-down hierarchical approach is the lack of a clear view of local congestion from the top hierarchies; final decisions may be made in some high-level hierarchy which may turn out to be unfavorable at lower levels, resulting in unremedied problems, as pointed out in [MS84, PT89, Lau87]. Adopting a Meet-in-the-Middle approach could help prevent

5.1. INTRODUCTION

such limitations. This approach is implemented in the hierarchical routing in Phase Three of the router.

3. *Vias should be avoided wherever straight-line paths can be made:*
   Vias correspond to contact openings in the chip fabrication process, which can lead to yield loss due to difficulties in opening contacts. In addition, they also consume more silicon area which lowers the chip's density. In global routing, more bends in the interconnection paths will likely produce more vias in the final layout. The proposed router will try to reduce the number of vias by first attempting to connect terminal nets with straight-line paths—in Phase One and Phase Two.

4. *Performance-driven layout issues need to be incorporated:*
   There has been an increasing need for performance and timing-driven layout systems [Kan90, JK89, Dai87]. The router fully incorporates performance considerations by associating a priority value to each net, which can be included as one component in the *benefit* function to be maximized by the 0-1 ILP in each of the routing phases. Thus, timing-critical nets are given preference and will be routed in the most optimal way possible.

The global routing algorithm is designed to use three layers of metallic interconnects for SOG arrays. The algorithm is subdivided into three distinct phases. Figure 5.1 provides an overview of the

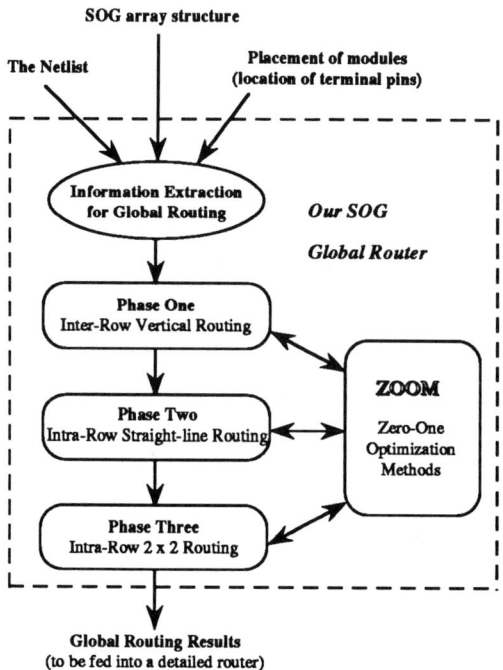

Figure 5.1: An overview of the SOG global router.

entire routing algorithm: the input information it accepts, the different routing phases and the interface with ZOOM [ZOO87], which is one of the well-established 0-1 ILP solvers available, and, finally, the output produced.

5.2 Extracting Global Routing Information

Before the SOG router actually performs the global routing of nets, information pertaining to the problem is extracted. During this process, a two-dimensional array of global routing cells (GRCs) is created

5.2. EXTRACTING GLOBAL ROUTING INFORMATION 147

using information from the netlist, the terminal pin locations that are determined during placement, information pertaining to the SOG array structure, and the preroutings or blockages in the macro-cells. Such extraction captures as much information of the chip image as possible, including the fact that SOG routing is being performed, while filtering out extraneous details.

In order to preserve the natural row structure of the SOG base array, the vertical pitch of the GRCs is determined by the number of horizontal tracks within each row of the SOG base array. The horizontal pitch of the GRCs is chosen by the user so that the total number of GRCs is kept to a manageable number. Since each n - p diffusion pair for CMOS circuits is explicitly represented by two subrows within a row of GRCs, the number of rows in the array is determined from the actual structure and size of the gate array to be used.

GRCs which contain at least one terminal pin of a particular net are designated as *terminal-GRCs*. The vertical and horizontal pitches of the GRCs determine the *boundary capacities* and the *via capacities* of the GRCs, which are then used in the 0-1 ILPs to specify the number of global routes allowed to cross each GRC boundary, and the number of global routes allowed to make bends within each GRC, respectively. Once the prerouting in the macro-cells is extracted, the *boundary* and *via capacities* of affected GRCs are appropriately reduced to model the blockages. As the routing algorithm creates new routes for the various nets, these capacity values are accordingly updated.

5.3 Global Routing Phases

Following the information extraction step, global routing is then performed using *three* layers of metal interconnects. Global routes for each net, composed of path segments which traverse a *subset* of the GRCs, are formed to connect the terminal pins of the net. For each of the nets, the route is decomposed into vertical and horizontal path *segments* instead of forming the entire Rectilinear Steiner Tree (RST) at once. As mentioned in Section 5.1, the construction of RSTs is NP-complete and hence highly time-consuming. These path segments are then routed separately in the three global routing phases until all of the pins are connected together.

In each routing phase, all of the nets are considered *simultaneously* instead of being routed in a particular net ordering. Possible path segments are identified and represented as 0-1 variables, following which 0-1 ILPs are formulated to select the optimal paths, with the objective of *maximizing* the *benefit* of selecting a set of paths. The constraints in these 0-1 ILPs are imposed by the circuit routing

5.3. GLOBAL ROUTING PHASES

features particularly important in SOG routing, namely, *over-the-cell* routing track availability and via placement availability. General circuit routing issues such as the avoidance of loop or cycle formation and the selection of at most one of many possible paths for each net are also considered. An external mathematical software package called *Zero-One Optimization Methods* (ZOOM) is used to perform the 0-1 ILP solution [ZOO87]. The solution obtained from ZOOM is used to decide how the nets are to be routed globally. This approach will allow any ILP package to be used. The three global routing phases, performed in the following order, are

Phase One (Interrow Routing)

A straight-line metal3 (topmost layer) vertical path segment, which will serve as the vertical *backbone* in the completed spanning tree, is formed for each net having terminals in *different* GRC rows.

Phase Two (Intrarow Straight-line Routing)

Straight-line path segments are formed within each GRC row in two subphases, namely, *Phase 2V*, in which only metal1 (bottommost layer) *vertical* path segments are formed between the two subrows of each GRC row, and *Phase 2H*, in which only metal2 *horizontal* path segments are formed within each subrow. These path segments can be thought of as *ribs* emanating from the metal3 backbone formed in Phase One. Although other variations of layer assignments may be possible, this particular arrangement was chosen to reduce the complexity.

Phase Three (Intrarow Two-by-two Routing)

Circuitous routes, which could *bend* around preroutings or blockages, are formed within each GRC row, using a hierarchy of 2×2 cell wiring problems. Path segments formed in this phase need not necessarily be confined within the rectangle bounding all of the terminal pins of a net.

5.3.1 Phase One Routing

In Phase One routing of the global router, INTERROW ROUTING is done by forming a straight-line vertical metal3 path segment for each net with terminal-GRCs in *different* GRC rows. This path will extend the entire *vertical span* of a net, which is defined as an ordered pair of integers, $[v, \lambda]$, where v is the number of the topmost GRC row containing a terminal-GRC of the net, and λ is the number of the bottommost GRC row with a terminal-GRC.

A net will be considered for Phase One routing only if $v \neq \lambda$ and no path segment is already formed. For such nets, possible vertical path segments are identified for each column in which the net possess a terminal. Each of these possible paths extends the entire vertical

5.3. GLOBAL ROUTING PHASES

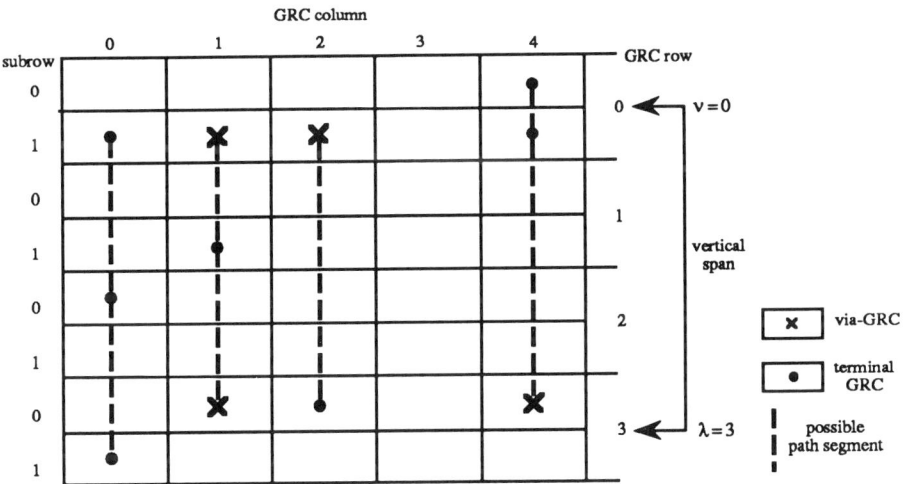

Figure 5.2: Possible vertical path segments for inter-row routing.

span of the net. If the starting and/or ending GRC of the selected path path is not a terminal-GRC, then the GRC becomes a *via-GRC*, which is defined as a GRC in which a via is formed by the global router. Figure 5.2 illustrates the possible vertical path segments for a sample net with a vertical span of $[0, 3]$.

A few word about notation are in order at this point. As mentioned earlier, the GRCs are arranged in a two-dimensional grid. Additionally, each GRC has two subrows. This is illustrated in Figure 5.2. The kth subrow of the GRC in position (i, j) in the two-dimensional grid is henceforth referred to as the (i, j, k)th GRC.

For each net, one of the possible path segments is chosen by the Phase One 0-1 ILP. The ILP is formulated as follows:

$$\text{MAXIMIZE} \sum_{\text{all } n} \sum_{\text{all } j} B_{n,j} \, P_{n,j}^{m3}$$

SUBJECT TO

Path Usage Constraints:
for each net, n,
$$\sum_{\text{all } j} P_{n,j}^{m3} \leq 1$$

South Boundary Constraints:
$\forall i, j, k$, for the (i,j,k)th GRC,
$$\sum_{\text{all } n_{i,j,k}} P_{n_{i,j,k},j}^{m3} \leq S_{i,j,k}^{m3}$$

Via Constraints:
$\forall i, j, k$, for the (i,j,k)th GRC,
$$\sum_{\text{all } n_{i,j,k}} P_{n_{i,j,k},j}^{m3} \leq V_{i,j,k}$$

AND *0-1 Variable Constraints:* $\quad P_{n,j}^{m3} = 0 \text{ or } 1$

where $\quad P_{n,j}^{m3}$ is a Boolean variable which is 1 only when the metal3 path of net n is in column j, and is 0 otherwise;
$S_{i,j,k}^{m3}$ is the metal3 south boundary capacity of the (i,j,k)th GRC;
$V_{i,j,k}$ is the via capacity of the (i,j,k)th GRC;
and $\quad B_{n,j}$ is the *benefit* value of partially connecting the net n using the path in column j.

The *benefit* value, $B_{n,j}$, of partially connecting net n using a vertical path segment in GRC column j, $P_{n,j}^{m3}$, is computed as follows:

$$B_{n,j} = P_n + T_{n,j} - X_{n,j} - \frac{R_{n,j}}{R_{total}} + 3,$$

5.3. GLOBAL ROUTING PHASES

where P_n : the priority value of net n, $0 \leq P_n \leq \mu$,
the default value of μ is 10;
$R_{n,j}$: the number of GRC rows spanned by the vertical path
R_{total} : the total number of GRC rows in the routing area;
$X_{n,j}$: the number of via-GRCs created at each extreme
end of the path, should it be used;
and $T_{n,j}$: the number of terminal-GRCs traversed by the path.

This expression for $B_{n,j}$ is based on a number of considerations, including the selection of 1) shorter paths over longer ones; 2) paths needing fewer vias; and 3) paths which traverse more terminal-GRCs. These considerations translate to the heuristic computation in the following way:

- component P_n is a *positive* measure of benefit. A path belonging to a net of a higher priority, P_n, should be given preference over another path belonging to a net with a lower priority.

- component $T_{n,j}$ is considered a *positive* measure of benefit. A path which traverse more terminal-GRCs is likely to be closer to more actual terminal pins; as such, it should be preferred over paths which traverses fewer terminal-GRCs.

- component $X_{n,j}$ is considered a *negative* benefit because it reflects the penalty of adding vias.

- component $R_{n,j}$ reflects the length of the vertical path. To discourage the use of long wires when shorter ones are available, the ratio, $\frac{R_{n,j}}{R_{total}}$, is *subtracted*.

- The *benefit* value of any line must be positive and nonzero to ensure that it is considered for routing. Since $\min(T_{n,j}) = 1$, $\max(X_{n,j}) = 2$ and $\max(\frac{R_{n,j}}{R_{total}}) = 1$, the constant component, 3, is *added* so that $\min(B_{n,j}) = 1$.

During this phase of the routing, the user can specify the amount of area, in terms of columns and rows, that should be considered for simultaneous routing. This feature allows the user to break up very large problems into smaller, more manageable problems. In effect, a crude partitioning of the circuit is performed, where no consideration is given to what is contained in each partition. With the use of this feature, the global router was capable of handling test cases of over 108,000 transistors and 9,400 nets.

5.3.2 Phase Two Routing

Since Phase One routing has already taken care of routing *between* different GRC rows, all of the interconnections created in Phase Two will be made *within* each GRC row. In this respect, each GRC row constitutes a separate subproblem. Moreover, only straight-line path segments will be formed. Hence, Phase Two routing involves IN-TRAROW STRAIGHT LINE ROUTING. It is divided into two subphases, namely, Phase 2V (Vertical) and Phase 2H (Horizontal) routings.

5.3. GLOBAL ROUTING PHASES

Phase 2V (Vertical) Routing

In Phase 2V, straight-line, vertical, metal1 interconnects are formed between pairs of terminal-GRCs in the two different subrows of a row. All such possible paths are identified. Figure 5.3 shows an example of a net with the two possible straight-line metal1 paths in a GRC row. Then, the 0-1 ILP for Phase 2V in a GRC row r is formulated to select *as many* of these paths as possible. The 0-1 ILP is formulated as follows:

$$\text{MAXIMIZE} \sum_{\text{all } n} \sum_{\text{all } j} B_{n,j} \, P_{n,j}^{m1}$$

SUBJECT TO
South Boundary Constraints:
$$\forall \text{ column } j \text{ (in row } r\text{)}, \qquad \sum_{\text{all } n_j} P_{n_j,j}^{m1} \leq S_{r,j,0}^{m1}$$

AND *0-1 Variable Constraints:* $P_{n,j}^{m1} = 0 \text{ or } 1$

where $P_{n,j}^{m1}$: a Boolean variable that is 1 only when the metal1 path of net n is in column j (of row r), and is zero otherwise;
$S_{r,j,0}^{m1}$: the metal1 south boundary capacity of the $(r,j,0)$th GRC;
and $B_{n,j}$: the *benefit* value of partially connecting the net n using its metal1 path in column j (of row r).

In the 0-1 ILP, a metal1 path of net n in GRC column j becomes a 0-1 variable, $P_{n,j}^{m1}$. Its associated *benefit* value, $B_{n,j}$, is computed as

$$B_{n,j} = P_n + 1,$$

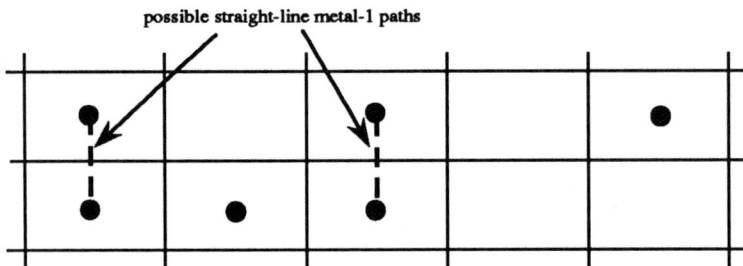

Figure 5.3: Possible straight-line metal1 paths.

where P_n is the priority value of net n, $0 \leq P_n \leq \mu = 10$, as in phase one. The constant, 1, is included so that $\min(B_{n,j}) = 1$. From this computation, it is obvious that the benefit value of the paths can differ only by the priority values of the net to which the path belongs; otherwise, all metal1 paths would be equally beneficial when they are selected. A path will be rejected only when it has to compete with another path belonging to a higher priority net for the same routing space. For Phase 2V of GRC row r, the only set of constraints is imposed by the metal1 south boundary capacity, $S^{m1}_{r,j,0}$, of the $(r, j, 0)$th GRC.

Phase 2H (Horizontal) Routing

In Phase 2H straight-line metal2 interconnects are formed between pairs of terminal-GRCs, via-GRCs and/or vertical metal3 paths in the *horizontal* direction. The possible straight-line metal2 paths of a net are shown in Figure 5.4.

5.3. GLOBAL ROUTING PHASES

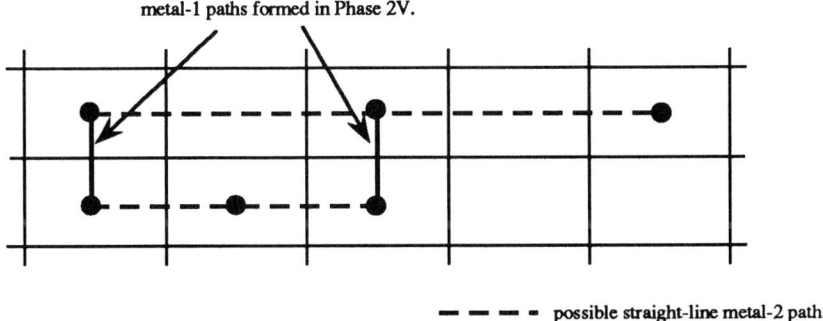

Figure 5.4: Possible straight-line metal2 paths.

The 0-1 ILP formulation for Phase 2H routing in GRC row r is given as:

$$\text{MAXIMIZE} \sum_{\text{all } n} \sum_{\text{all } k} \sum_{\text{all } j} B_{n,k,j} \, P^{m2}_{n,k,j}$$

SUBJECT TO

East Boundary Constraints:
$\forall j, k$, (in row r),
$$\sum_{\text{all } n_{r,j,k}} P^{m2}_{n_{r,j,k},j} \leq E^{m2}_{r,j,k}$$

Via Constraints:
$\forall j, k$, (in row r),
$$\sum_{\text{all } n_{r,j,k}} P^{m2}_{n_{r,j,k},j} \leq V_{r,j,k}$$

Loop Avoidance Constraints:
for each loop, l,
$$\sum_{\text{all } k} \sum_{\text{all } j_l} P^{m2}_{n_l,k,j_l} \leq (N_l - 1)$$

AND *0-1 Variable Constraints:* $P^{m2}_{n,k,j} = 0$ or 1

where $P^{m2}_{n,k,j}$: a Boolean variable that is 1 only when the horizontal metal2 path of net n in subrow k (of row r), starting at column j, and is zero otherwise;

$E^{m2}_{r,j,k}$: the metal2 east boundary capacity

of the (r,j,k)th GRC;
$V_{r,j,k}$: the via capacity of the (r,j,k)th GRC;
N_l : the total number of possible horizontal metal2 paths for net n in loop l;
and $B_{n,k,j}$: the *benefit* value of partially connecting the net n using its horizontal metal2 path in subrow k (of row r), starting at column j.

In the above 0-1 ILP, the *benefit* value, $B_{n,k,j}$, associated with each path is computed heuristically as

$$B_{n,k,j} = P_n + \min\left(\frac{C_{total}}{C_{n,k,j}}, 10\right),$$

where P_n : the priority value of net n, $0 \leq P_n \leq \mu = 10$;
$C_{n,k,j}$: the number of GRC columns spanned by the horizontal path;
and C_{total} : the total number of GRC columns in the routing area.

The two components in this computation are both *positive* measures of benefit. As in previous routing phases, the priority of net n, P_n, is added to give preference to a path belonging to a net with a higher priority. The second component, $\min(\frac{C_{total}}{C_{n,k,j}}, 10)$, is intended to range in value from 1 to 10, so that it carries at most as much *weight* as P_n ($=10$), and at the same time, $\min(B_{n,j,k}) = 1$. Its derivation is thus explained: The term, $\frac{C_{total}}{C_{n,k,j}}$, is *inversely* proportional to the length of the horizontal path, $P_{n,k,j}^{m2}$. It reflects the preference for shorter paths, and has values ranging from 1 to C_{total} because $1 \leq C_{n,k,j} \leq C_{total}$. The lower limit automatically satisfies that of $\min(\frac{C_{total}}{C_{n,k,j}}, 10)$, while

5.3. GLOBAL ROUTING PHASES

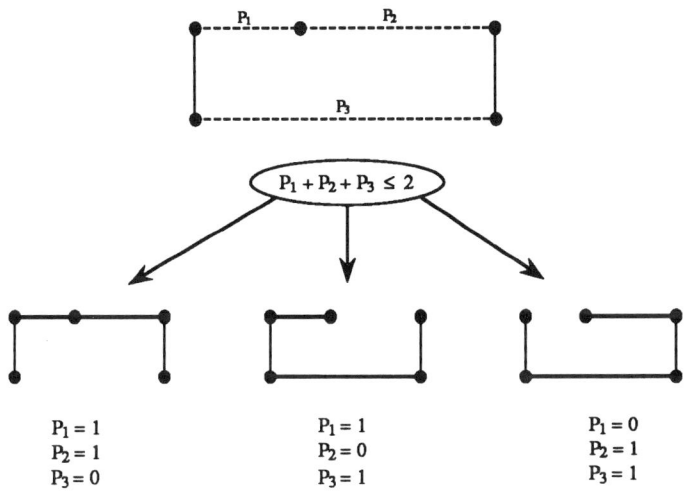

Figure 5.5: An example of a loop avoidance constraint.

the upper limit is forced to 10 by comparing it with the constant, 10, and taking the minimum of the two values.

It can be seen, e.g., in Figure 5.4, that loops within the same GRC row may form if some of the paths are simultaneously chosen. A loop, l, in a net, n, is formed if all of the N_l possible horizontal paths, $P^{m2}_{n_l,k,j_l}$, between two GRC columns with existing vertical path segments are selected by the 0-1 ILP. Such loops are avoided by forcing the 0-1 ILP to select only at most $N_l - 1$ paths by imposing loop avoidance constraints:

$$\text{for each loop, } l, \quad \sum_{\text{all } k} \sum_{\text{all } j_l} P^{m2}_{n_l,k,j_l} \leq (N_l - 1).$$

For example, in Figure 5.5, only two of the three possible paths will be selected, resulting in one of the three configurations given, thereby avoiding the formation of a loop connection.

5.3.3 Phase Three Routing

In Phase Three routing, a TWO-BY-N HIERARCHICAL GRID CELL APPROACH ($2 \times N$ routing) is used for intra-row horizontal metal2 interconnect and localized vertical metal1 interconnect, where N is the number of GRC columns. Like Phase Two routing, all path segments are formed *within* each GRC row. The goal of this routing phase is to complete the routings left undone after all possible *straight-line* path segments have been made in the previous phases. As opposed to the purely straight-line path segments made in Phase One and Phase Two, circuitous routes which take as many *bends* as they need to avoid the preroutings in the modules or gates are made in this phase. Each route is a *subtree* of a graph formed by associating a vertex with each of the $2N$ GRCs in a GRC row and an edge connecting adjacent pairs of vertices (see Figure 5.6). It consists of vertical and/or horizontal path segments connecting the terminal-GRCs of its net, forming via-GRCs at each bend. These path segments will be selected by 0-1 ILPs which take into consideration the GRC boundary and via constraints.

Phase Three routing employs an adaptation and extension of the

5.3. GLOBAL ROUTING PHASES

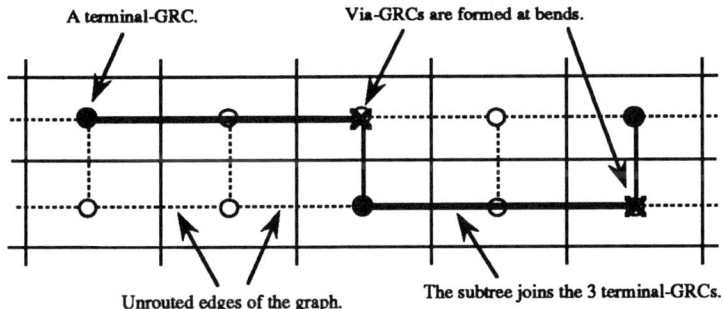

Figure 5.6: A possible route to be formed in Phase Three routing.

$2 \times N$ routing methodology by Burstein and Pelavin [BP83]. Each cell in the 2×2 cell wiring problems is formed by grouping one or more GRCs in the same subrow of GRCs, and is referred to as a *Macro-GRC*. The hierarchical setup of Phase Three routing is very similar to that employed in [BP83]. However, there are a few important differences between the two routing schemes, as summarized in Table 5.1.

Table 5.1: Differences between [BP83] and Phase Three routing.

Burstein & Pelavin's Method	Phase Three of the SOG router
Purely $2 \times N$ routing — not preceded by straight-line interconnections.	$2 \times N$ routing is preceded by straight-line interconnections in Phase One and Phase Two.
No corrective measure was taken to level out nonuniform boundary capacities.	Phase 2V & 2H serve to level out non-uniform GRC boundaries.
Restrictive routing — a wire route cannot cross vertical boundary twice.	Nonrestrictive routing — a route may cross a vertical boundary twice.
Fewer routing configurations.	More routing configurations arising from connections made in Phase One and Phase Two.
Strictly top-down hierarchical approach.	A meet-in-the-middle approach is used, alternating between top-down and bottom-up routing.
Determines only the number of nets to be used for each configuration.	Determines the exact net usage, i.e., how each net is routed.

5.3. GLOBAL ROUTING PHASES

Phase Three routing is done in a combination of top-down and bottom-up hierarchical fashion, involving four steps as stipulated in the following algorithm:

For each row in the routing area, do {
  * Step 1:
    Invoke the algorithm to perform 2 × 2 macro-GRC routing
    at the lowest hierarchy (level == L);
  * Step 2:
    Perform *Meet-in-the-Middle* routing, as follows:
      - Initialize:
        top-level = 1; bottom-level = L − 1; TOP = 0 (FALSE);
      - While (top-level ≤ bottom-level), i.e., if top and bottom
      levels have not met, do {
        Alternate between top-down and bottom-up routing,
        as follows:
          TOP = (TOP + 1) modulo 2; /* toggle TOP flag */
          If TOP,
            do TOP-DOWN routing, as follows:
            - Invoke the algorithm to perform 2 × 2
            macro-GRC routing at level == top-level;
            - Move down one level: top-level = top-level + 1;
          else
            do BOTTOM-UP routing, as follows:
            - Invoke the algorithm to perform 2 × 2
              macro-GRC routing at level == bottom-level;
            - Move up one level:
            bottom-level = bottom-level − 1;
      }
  * Step 3:
    Perform *Downward-Direction* 2 × 2 macro-GRC routing at
      each level, starting from one level below which
      TOP-DOWN routing stopped, to one level above the lowest
      hierarchy (level == L − 1);
  * Step 4:
    Invoke the algorithm to perform 2 × 2 macro-GRC routing at
      the lowest hierarchy (level == L);
}

A space-time diagram for a 2 × 16 problem is given in Figure 5.7 to illustrate the hierarchical steps involved in Phase Three routing.

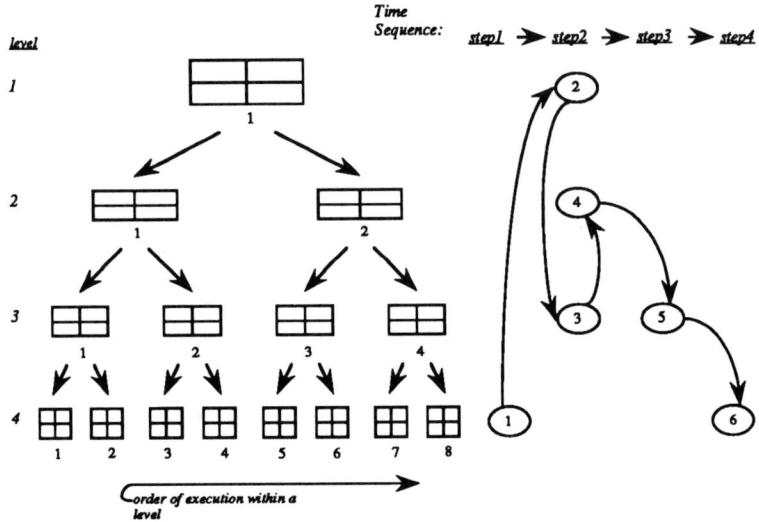

Figure 5.7: Steps in solving a 2 × 16 problem.

The most basic unit of Phase Three routing is 2 × 2 Macro-GRC routing. The formation of Macro-GRCs in the 2 × 2 problem necessitates the following definitions:

- A *macro-node* is defined as a macro-GRC in which at least one of its member GRCs is a terminal-GRC, a via-GRC or a *potential* via-GRC, which could be formed in all levels *except* the lowest of the hierarchy. A *potential* via-GRC, illustrated in Figure 5.8 is one in which a horizontal path segment crossing a macro-GRC boundary ends, and a decision to make a horizontal or vertical segment from this GRC is pending until routing

5.3. GLOBAL ROUTING PHASES

in the lowest hierarchy is done.

- A *macro-edge* is formed when there is at least one routed segment crossing the boundary between two macro-GRCs.

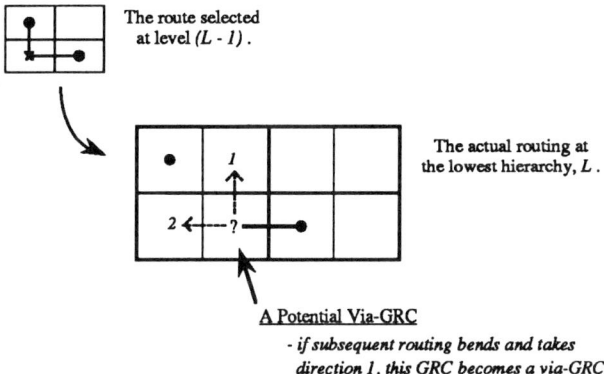

Figure 5.8: A potential via-GRC.

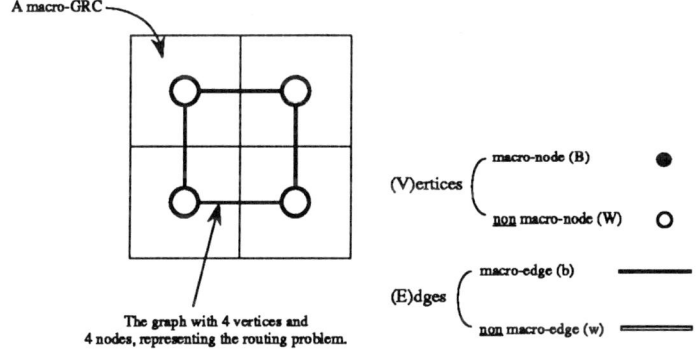

Figure 5.9: Representation of the 2×2 macro-GRC routing problem.

The goal of 2×2 macro-GRC routing is to classify the partially connected nets, identify all possible macro-edges between the macro-nodes and select some or all of these edges to achieve an optimal routing. This problem is represented as a graph with 4 vertices (V) and 4 edges (E), with one vertex in each macro-GRC and one edge crossing each cutline between two macro-GRCs (see Figure 5.9). A vertex is black (B) if it is in a macro-node; otherwise, it is white (W). Similarly, a black (b) edge represents a macro-edge, while an invisible or white (w) edge indicates the absence of a macro-edge.

Polya's Enumeration Method [Tuc84] is used to determine the number of different *equivalence classes* of a 2×2 net configuration with e macro-edges and n macro-nodes, where $e = 0, 1, 2, 3, 4$ and $n = 0, 1, 2, 3, 4$. There are a total of 70 equivalence classes. Fortunately, all 70 equivalence classes need not be considered in the 2×2 routing problem because many of these classes are either physically impossible net configurations or configurations in which no further routing is needed. Details of all of the computations and results can be found in [Lek90]. In the final analysis, only 29 equivalence classes have to be considered for routing in Phase Three. This number, however, is still significantly larger than the 13 configurations in [BP83]. The nets in the 2×2 macro-GRC routing problems will be classified into these 29 classes. Once the net classification is done, the possible routes for that particular net classification are then identified from a database containing all of the net configurations and alternative path connections, which are given in Figures 5.10 through 5.13. In other words, the identification of the possible routes in Phase Three is a fast and simple *table lookup* for the appropriate entry in the database of net classifications and routings.

5.3. GLOBAL ROUTING PHASES 167

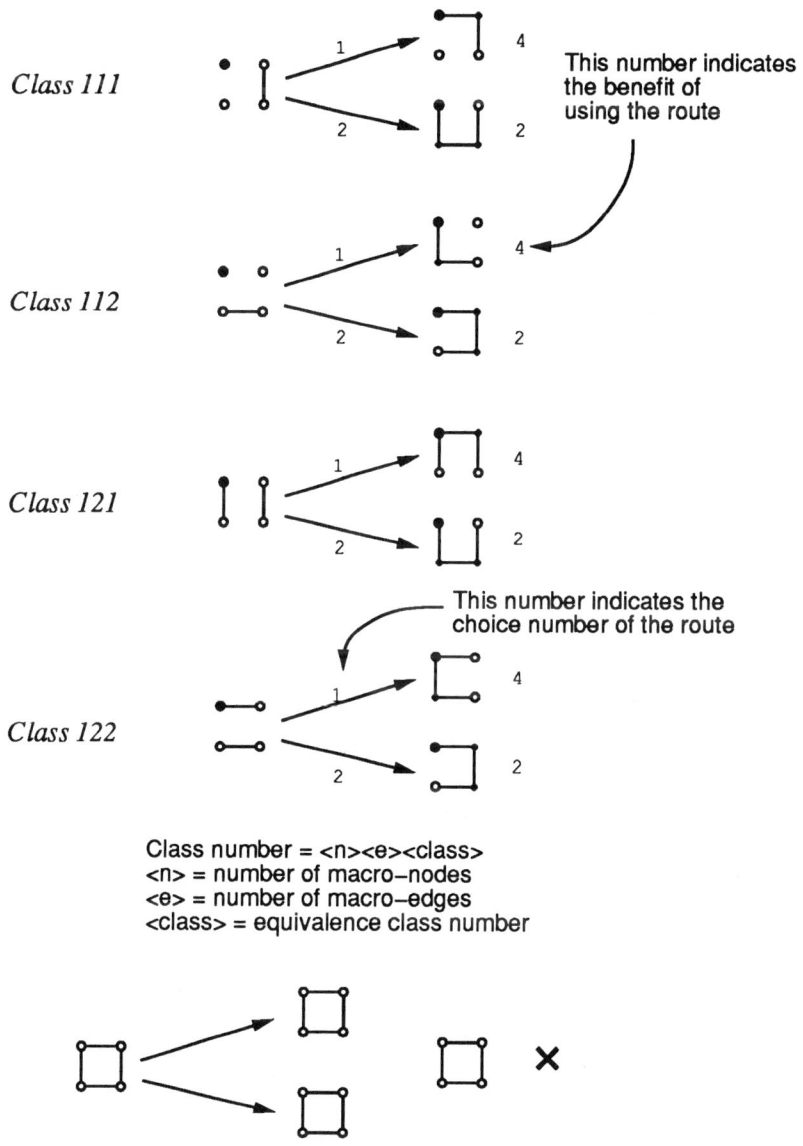

Figure 5.10: Possible routes in nets with one macro-node.

The 0-1 ILP formulated to select the possible routes for 2×2 macro-GRC routing is given as follows:

$$\text{MAXIMIZE} \sum_{\text{all } n} \sum_{\text{all } c} B_{n,c} R_{n,c}$$

SUBJECT TO

Net Connection Constraints:
for each net, n, $\quad \sum_{\text{all } c} R_{n,c} \leq 1$

Boundary Constraints:
for east boundary, $E(0)$, $\quad \sum_{\text{all } n_{E(0)}} R_{n_{E(0)},c} \leq E^{m2}_{r,j',0}$

for east boundary, $E(1)$, $\quad \sum_{\text{all } n_{E(1)}} R_{n_{E(1)},c} \leq E^{m2}_{r,j',1}$

for south boundary, $S(0)$, $\quad \sum_{\text{all } n_{S(0)}} R_{n_{S(0)},c} \leq \sigma^{m1}_{S(0)}$

for south boundary, $S(1)$, $\quad \sum_{\text{all } n_{S(1)}} R_{n_{S(1)},c} \leq \sigma^{m1}_{S(1)}$

Via Constraints:
for each macro-GRC (x,y), $\quad \sum_{\text{all } n_{(x,y)}} R_{n_{(x,y)},c} \leq \vartheta_{(x,y)}$

AND *0-1 Variable Constraints:* $\quad R_{n,c} = 0$ or 1

where $R_{n,c}$: a Boolean variable that is 1 only for the choice c route for connecting net n, and is 0 otherwise;

$E^{m2}_{r,j',k}$: the metal2 east boundary capacity of the (r,j',k)th GRC, which is the rightmost GRC constituting the macro-GRC $(k,0)$;

$\sigma^{m1}_{S(k)}$: the *average* value of the metal1 south boundary capacity of all of the GRCs constituting the macro-GRC $(0,k)$;

$\vartheta_{x,y}$: the *average* value of the via capacity of all of the GRCs constituting the macro-GRC (x,y);

and $B_{n,c}$: the *benefit* value of partially connecting the net n using its choice c path.

5.3. GLOBAL ROUTING PHASES

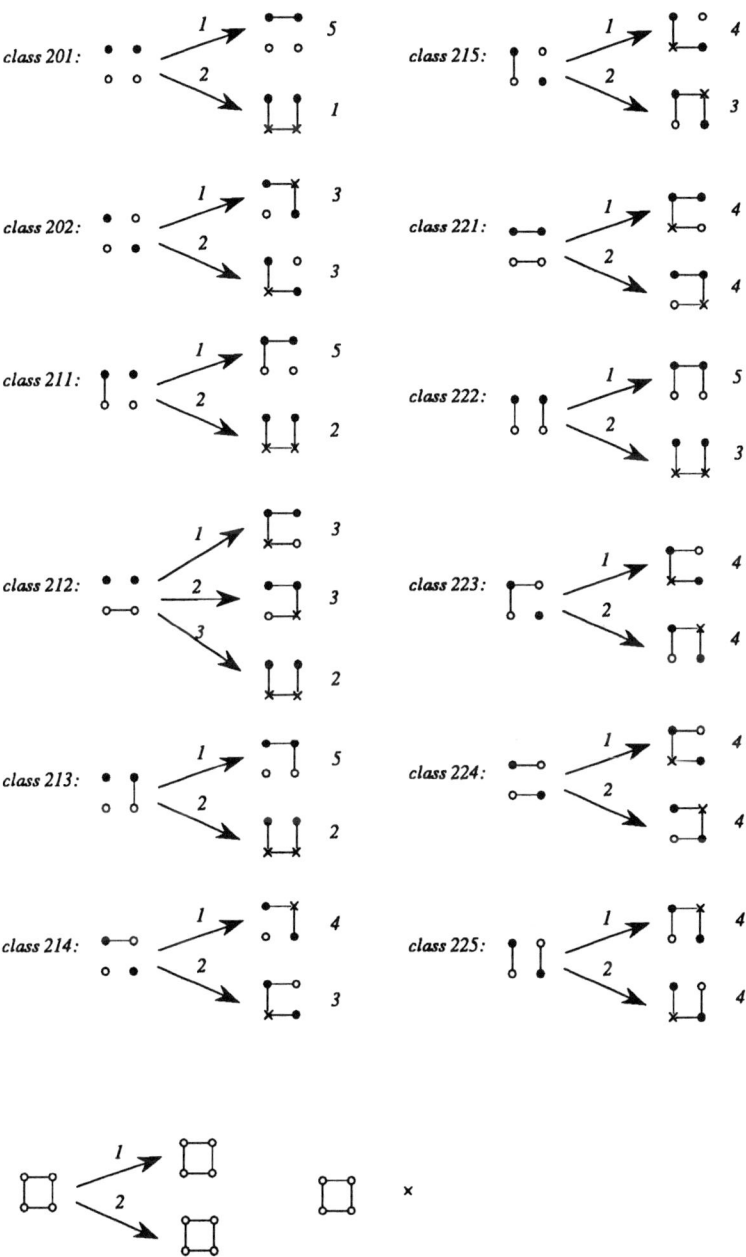

Figure 5.11: Possible routes in nets with two macro-nodes.

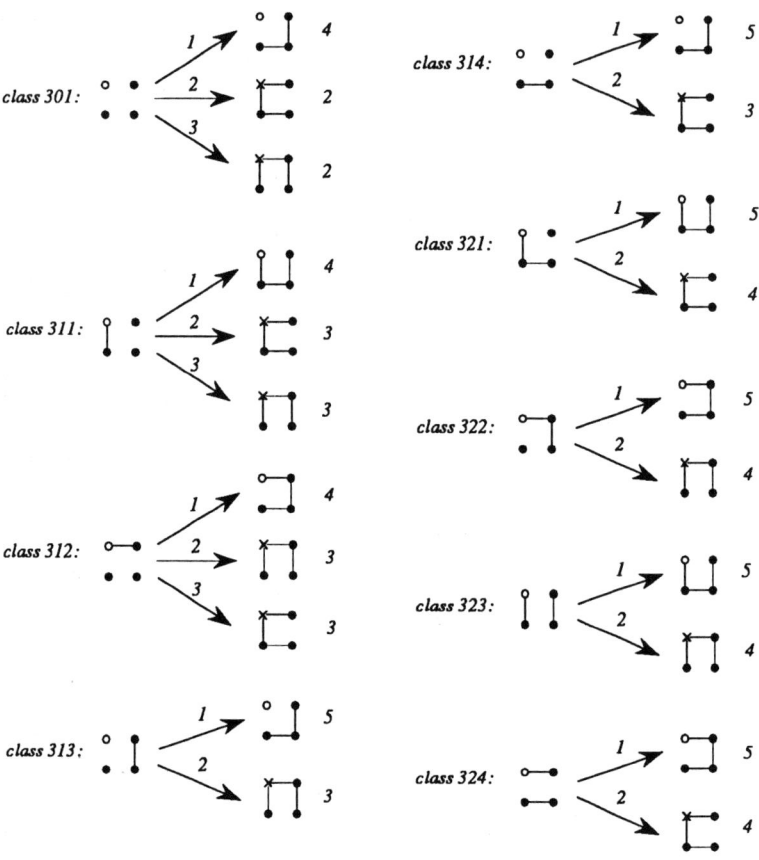

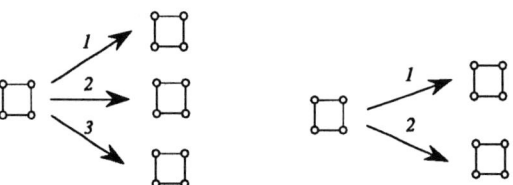

Figure 5.12: Possible routes in nets with three macro-nodes.

5.3. GLOBAL ROUTING PHASES

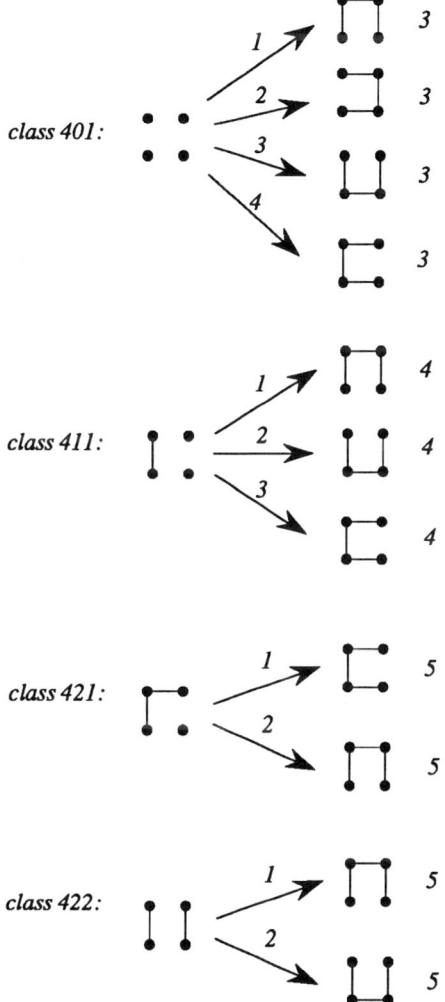

Figure 5.13: Possible routes in nets with four macro-nodes.

In the Phase Three 0-1 ILP formulation, as shown in Figure 5.14, the boundary capacities for $E(0)$ and $E(1)$ are *actual* GRC boundary capacities because the horizontal path segments are actually crossing the east boundary of the (r, j', k)th GRC. On the other hand, the *average* of all of the GRC south boundary capacities constituting the macro-GRC $(0, k)$ is used as a composite index of vertical track availability in $S(0)$ and $S(1)$ since no final decision as to exactly which vertical path segment should cross which GRC south boundary is made until the lowest level of the hierarchy.

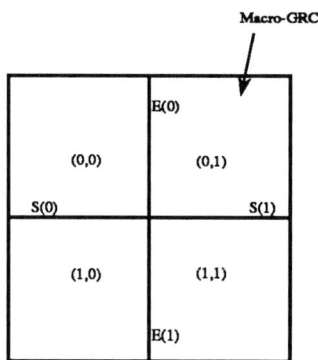

Figure 5.14: Boundary and via capacities of 2×2 macro-GRCs.

The benefit value, $B_{n,c}$, of connecting net n using the choice c route, $R_{n,c}$, in the 2×2 routing problem is determined from the following expression:

$$B_{n,c} = P_n + \{6 - F_{n,c}\},$$

where P_n : priority value of net n, $0 \leq P_n \leq \mu = 10$;
and $F_{n,c}$: total number of possible vias *and* path segments added if choice c is used to connect net n, $1 \leq F_{n,c} \leq 5$.

This computation is the sum of two components. The first component is the same priority value, P_n, of net n used in previous routing phases. The second component, $\{6 - F_{n,c}\}$, gives preference to routes which require fewer path segments (i.e., shorter wire length) and vias to fully connect the nets. All of the route configurations give rise to at most five path segments and vias combined, that is, $\max(F_{n,c}) = 5$. As such, the constant, 6, is included so that the values of the second component range from 1 to 5, and that $\min(B_{n,c}) = 1$. The second component of the benefit value for each of the possible routes in all of the net configurations is shown in Figures 5.10 through 5.13.

5.4 Global Routing on Medium-sized Arrays

The global router for sea-of-gates and custom logic arrays has been implemented in C and runs on a UNIX operating system linked to an external LP solver, ZOOM. For illustration of the routing methods, consider a small random circuit with 10 nets, called **i10**. The nets reside in a routing area with 3 GRC rows and 8 GRC columns. The number of rows was determined by the actual layout of the SOG array, while the number of columns was chosen to keep the number of GRCs small. The routing result is shown in Figure 5.15. To trace through the three routing phases, one of the nets, net G, is taken as an example. Figure 5.16 shows the location of net G's terminal-GRCs before the global routing, the partial routes formed after Phase One and Phase Two, as well as the final path found after Phase Three. In Phase One routing, the possible paths of net G exist in GRC columns 3, 5 and 7, spanning from GRC row 0 to GRC row 2. The vertical metal3 path in GRC column 7 was selected by the Phase One 0-1 ILP since it is the shortest path which connects 2 terminal-GRCs and spans rows 0 to 2. Subsequently, a metal1 vertical path and a metal2 horizontal path were formed in GRC row 2 during Phase 2V and Phase 2H, respectively. At this stage, four out of five terminal-GRCs of net G have already been connected together. Finally, in Phase Three routing, the fifth terminal-GRC was linked to the partial route by an inverted-L path consisting of a horizontal metal2 segment and a vertical metal1 segment joined at a via created in GRC $(0,7,0)$. Thus, global routing in net G is completed. The other nine nets in circuit **i10** are routed in the same manner.

Test cases were then run to observe the effect on the net completion rate for a varying number of tracks. Figure 5.17 shows the percentage of nets completely routed for various numbers of available horizontal and vertical tracks. The results show that the global router is capable of routing 100 percent of the nets provided a suf-

5.4. GLOBAL ROUTING ON MEDIUM-SIZED ARRAYS   175

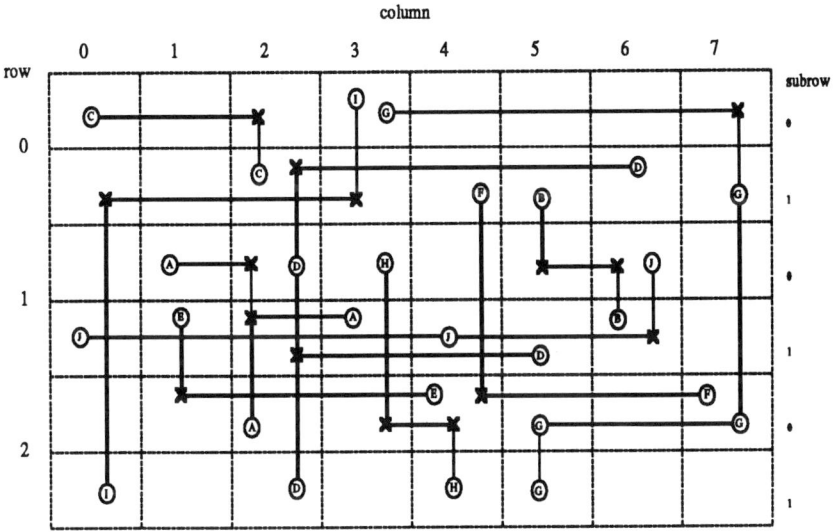

Figure 5.15: Routing area showing the routing result of circuit **i10**.

ficient number of tracks is available. Once the number of tracks - either horizontal or vertical - falls below a certain cut-off point, the net completion percentage drops drastically. The cut-off point is the physical minimum required for the routing to be performed successfully on all of the nets. The factors that influence the cut-off point include the location of net terminal, the number of nets, prerouting

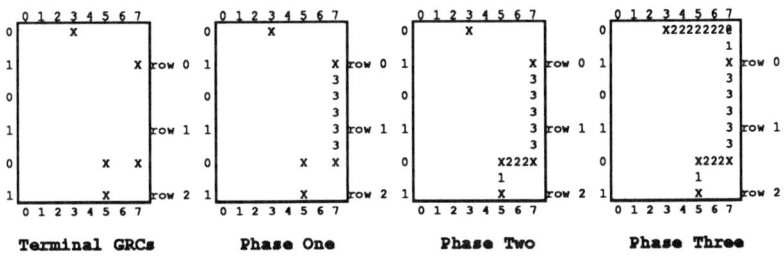

Figure 5.16: The routing process for net G of circuit **i10**.

blockages, and the local density of nets.

The problem of large blockages in the circuit can be thought of as locally decreasing the number of tracks available. If the number of tracks available is larger than that at the cut-off point, the blockage will have no effect on the net completion rate. However, if the number of tracks is lower than that at the cut-off point, two possibilities exist. In one case, the blockage will occur in a heavily congested area of the design. For this case, the blockage will cause a decrease in the completion percentage. If the blockage is not in a heavily routed area, the effect of the completion ratio is dependent on how large the blockage is compared to the minimum number of tracks needed in that area of the design.

The program runs successfully on several industrial circuits. The results for a CMOS circuit, **c162**, and a BiCMOS circuit, **c1318**, are presented here. These circuits consist of gate modules which are built based on the sea-of-gates base cell array in [Gal90]. Placement of the modules in these circuits was done using *TimberWolfSCv5.4* [Tim87], with the *row separation* parameter of the latter program set to 0 so as to create truly *channelless* designs. Table 5.2 shows all of the information pertaining to the type, size and placement of the circuits. Also listed are the packing densities of these circuits within which all global routings were to be confined.

5.4. GLOBAL ROUTING ON MEDIUM-SIZED ARRAYS

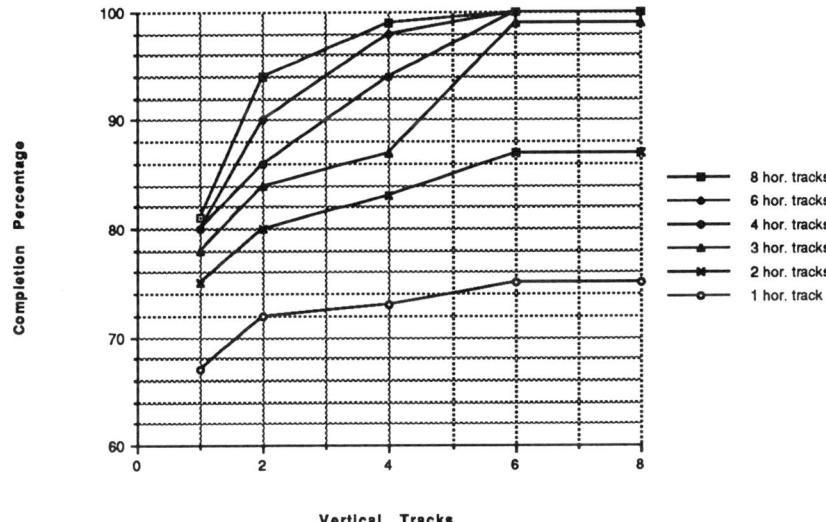

Figure 5.17: Percentage of nets routed vs. available horizontal and vertical tracks.

Table 5.2: Type, size and placement information of **c162** and **c1318**.

Name of Circuit	**c162**	**c1318**
Type of Circuit	CMOS	BiCMOS
No. of Transistors	450	4657
No. of Modules	116	956
No. of Nets	162	980
Layout/Routing Area	34,145 μm^2	724,992 μm^2
Packing Density	75.9 μm^2/trans.	155.7 μm^2/trans.

The program was then run on each of the circuits with varying GRC pitch sizes. Table 5.3 gives the global routing statistics for the two circuits. It can be seen that both the number of 0-1 ILPs solved and the corresponding CPU-runtimes for the global routing increase as the number of GRC columns is increased to obtain finer global routes using smaller GRCs, i.e., more GRC columns. It also demonstrates the flexibility of the method and allows the choice of an appropriate GRC size that compromises the speed and the quality of the global routing result. In all cases, all of the nets were *completely* routed globally *within* the layout area given in Table 5.2; the program achieved 100 percent *over-the-cell* global routing for both **c162** and **c1318** without having to introduce routing channels or overflow existing routing tracks.

The packing densities of the circuits compare very favorably with those of many SOG chips reported in the literature (see Figure 5.18 and Figure 5.19). The superior packing densities may have resulted from the superior placement and the ability to route all of the nets within this compact area without dedicated routing channels. Unfortunately, the exact reasons for improvement are difficult to access since design details of the other arrays are not available. Figure 5.18 compares the packing density of **c162** (its value is normalized at different feature sizes of the process technology and drawn as a dotted line) with those for CMOS chips from Fujitsu [SMN88], Hitachi [Tak89], Mitsubishi [Oka89], Philips [VEH90] and Stanford University [GH90]. Similarly, the packing density of **c1318** is compared with that for BiCMOS chips from Fujitsu [EST90], NTT LSI Laboratories [Yos89], and LSI Logic Corporation [Won88].

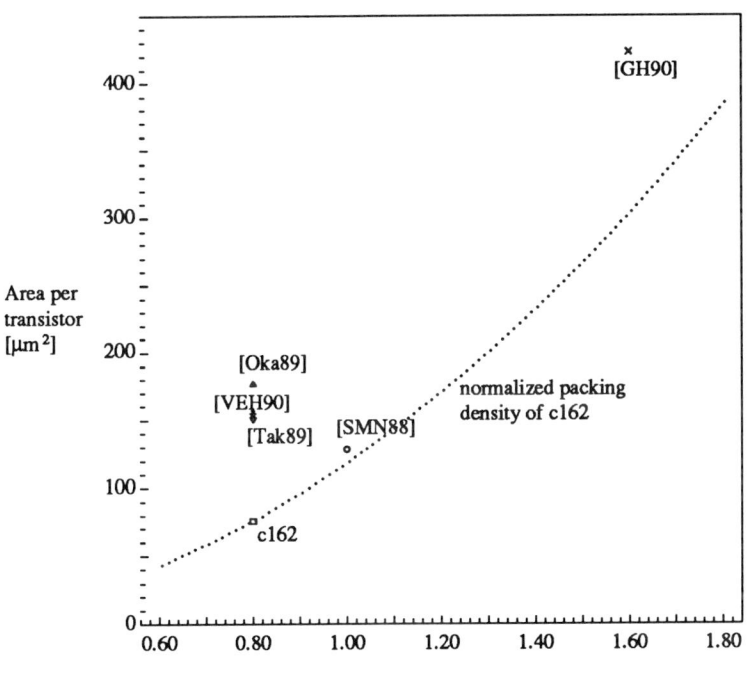

Figure 5.18: Packing densities of **c162** and reported CMOS chips.

5.5 Application to Custom Logic Layout

Next, an application of the ILP approach to custom logic module design is considered. Such a system can be used to generate com-

Table 5.3: Global routing statistics for circuits, **c162** and **c1318**.

Name of Circuit	c162	
No. of GRC rows	4	4
No. of GRC columns	8	16
P^h	7	4
P^v	9	9
No. of Nets Routed	162/162	162/162
No. of 0-1 ILPs (Actual/Max.)		
Phase One	1/8	2/16
Phase Two	5/8	5/8
Phase Three	22/52	25/108
Total	28/68	32/132
Convex Cpu-time	9.19 sec	16.00 sec

Name of Circuit	c1318		
No. of GRC rows	8	8	8
No. of GRC columns	8	16	32
P^h	73	36	18
P^v	9	9	9
No. of Nets Routed	980/980	980/980	980/980
No. of 0-1 ILPs (Actual/Max.)			
Phase One	8/8	8/16	16/32
Phase Two	16/16	16/16	16/16
Phase Three	72/104	127/216	379/472
Total	96/128	151/248	411/520
Convex Cpu-time	105.88 sec	204.13 sec	733.05 sec

5.5. APPLICATION TO CUSTOM LOGIC LAYOUT

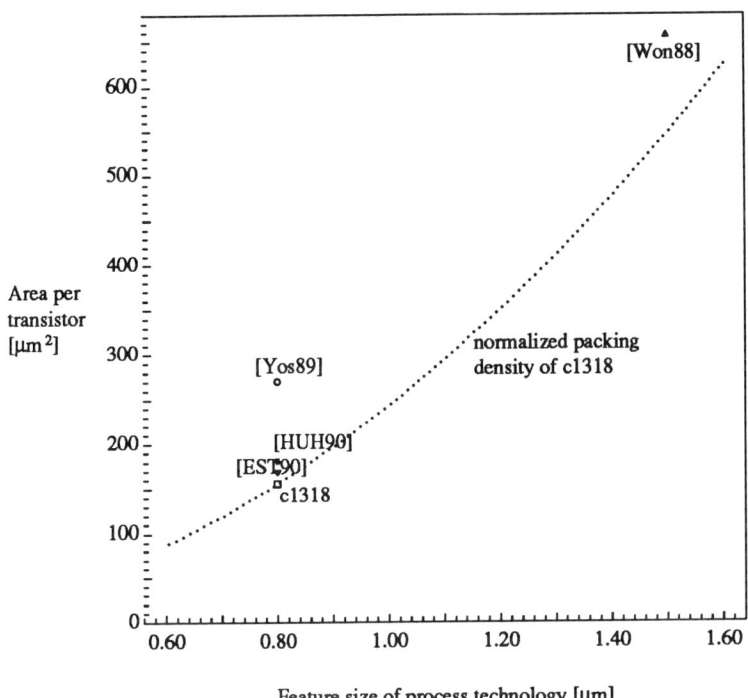

Figure 5.19: Packing densities of **c1318** and reported BiCMOS chips.

pact layout for logic blocks of around 1000 transistors from a given netlist. The system can combine a standard cell approach to placement with a sea-of-gates array layout to perform the routing and transistor level placement. The size of each cell will be an estimate based on the number of transistors within the cell. Initially, all transistors are assumed to be minimum width. Once the estimation is finished, a standard cell placement program, like TimberWolf, is used to perform the placement of the logic cells. This placement can then be mapped onto a sea-of-gates array for the rest of the process. The location of net terminals will be extracted from the netlist and placement of logic cells. The global routing program is used to determine which GRCs a particular net will pass through. Once the global routing phase is done, the fine routing and placement program will take the data formed from the previous stages of the process. The data from the global routing initially serves to create a set of loose, or unassigned, boundary constraints. As the program proceeds to process GRCs, some of the boundary conditions will change from being loose to tight, or restricted, boundary constraints. After the placement and fine routing is accomplished, the signal delay can be computed. A transistor sizing program can then be run to optimize the individual transistor sizes for speed and performance. The results of the transistor sizing will then update the netlist, and the process can be repeated until the chip design is completed. Figure 5.20 shows the global routing result for a four-bit adder circuit with 144 transistors and 45 nets. The circuit consisted of 36 NAND2 gates which were first placed with TimberWolf and then globally routed with this program.

5.6 Handling Very Large Circuits

A major problem in applying previous linear programming approaches to global routing problems has been due to the limited *size* of the cir-

5.6. HANDLING VERY LARGE CIRCUITS

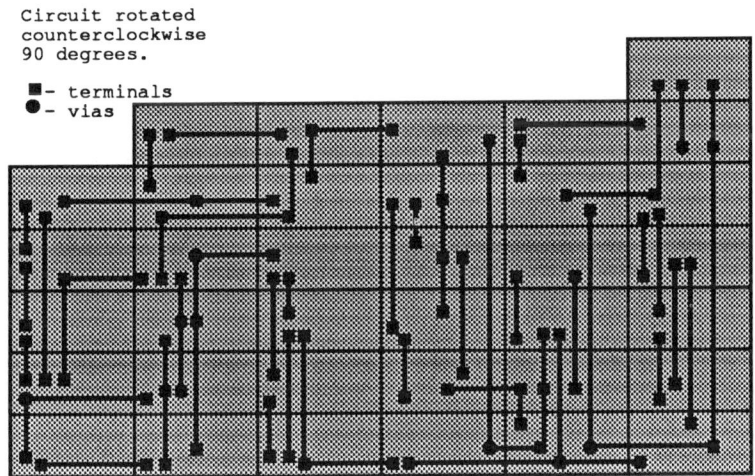

Figure 5.20: Placement and global routing of a four-bit adder.

cuits manageable by those methods. To demonstrate the capability of the SOG global router to handle larger SOG circuits, an industrial BiCMOS circuit, **c2396** containing over 27,000 transistors, was run using that particular company's SOG array structure. The placement of the modules *without* any channels between the array rows was done using *TimberWolfSC v5.4*. Subsequently, the program was able to route all 2396 nets within the channelless layout area with a CPU runtime of only 821.72 sec. The packing density achieved was 151.3 μm^2 per transistor, once again surpassing those of BiCMOS chips in [EST90, Yos89, Won88, HUH90]. The statistics are given in Table 5.4. To our knowledge, ILP handling of such a large circuit has not been reported.

Table 5.4: Global routing statistics for **c2396**.

Name of Circuit	**c2396**
Type of Circuit	BiCMOS
No. of Modules	3120
No. of Nets	2396
Layout/Routing Area	4,119,552 μm^2
No. of Transistors	27,220
Packing Density	151.3 μm^2/trans.
No. of GRC rows	8
No. of GRC columns	16
P^h	52
P^v	36
No. of Nets Routed	2396/2396
No. of 0-1 ILPs (Actual/Max.)	
Phase One	16/16
Phase Two	16/16
Phase Three	214/216
Total	246/248
Convex Cpu-time	821.72 sec

In another run, the 27,000-transistor circuit was duplicated and placed side by side. Connections between the two circuits were formed to make a circuit with 54,000 transistors and 4794 nets. Again, the program was able to route all of the nets within the allotted space and tracks. Experimental results for even larger circuit cases containing 108,000 transistors will be presented in the next section on runtime complexity. Currently, the maximum size of the circuits that can be handled by this method is limited by size of the

5.7. RUNTIME COMPLEXITY

problem the ILP solver can handle. Alternatively, the size limit problem can be overcome with circuit partitioning. A suitable block partitioning algorithm with *interface-pin assignment* was implemented in SOGR [PT89]. After block partitioning, the global routing method can be applied to each of the smaller circuit blocks, treating the interface pins as terminal pins of nets which cross the block boundary. The pin assignment step can be bypassed if additional metal layers become available to route all such nets.

The global router can be easily adapted to a parallel processing environment. Since this global routing approach has a row-based nature and the algorithms in Phase Two and Phase Three contain much inherent parallelism, parallel processors can be used to handle routings concurrently in separate GRC rows within a sub-block and also to handle parallel processing among different sub-blocks, thereby shortening runtimes for solving large problems.

5.7 Runtime Complexity

Time complexity tests were performed on the global routing program for various size test cases. To perform the test, the **c1318** circuit with 4657 transistors and 980 nets was duplicated several times with each net given a new name in each duplicate. These duplicates were placed next to each other to form test cases of 2, 4, 6, and 9 times the size of the original circuit. Connections were then made between the duplicates in 5% of all of the nets. This experimental set up has been suggested by both industrial colleagues and reviewers to overcome the problems in obtaining large industrial data. Global routing was performed on the four new test cases in addition to the original case on a Convex machine. Each run of the program was timed with the shell command *time*, and the results are shown in Figure 5.21. As the data shows, the increase in system cpu-time is approximately linear with the problem size. The overriding factor in the amount of time needed for any particular run is the number of 0-1 ILPs that are generated and solved. This number depends on the local density of nets, the number of rows and columns in each partition, the size of the overall circuit, and the problem size that the 0-1 ILP solver can handle.

To see if linear runtime complexity holds for very large circuits, the **c2396** circuit with over 27,000 transistors was duplicated in the same manner as described above. Again, the runtime complexity for system cpu-time was linear with the problem size up to a 108,000 transistor test case, as shown in Figure 5.21. It is interesting to note that these "artificial" circuits created by duplication seem to provide an over-estimate of the required runtime. As the figure shows, the runtime needed for the duplicated 4.6K transistor test case with 27K

5.7. RUNTIME COMPLEXITY

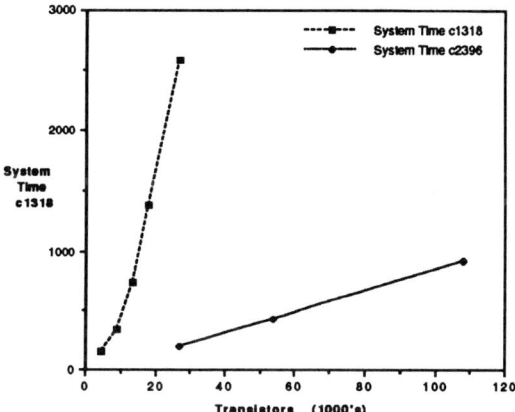

Figure 5.21: Runtime complexity of the global router.

transistors was greater than the time needed to run the single **c2396** case with 27K transistors. The reason for this overestimate stems from the fact that as the transistor count is artificially increased, the number of nets increases linearly with the number of transistors. Thus, even though the artificial test case with six **c1318** had about the same number of transistors as the single **c2396**, the number of nets in the first case was about 2.5 times as many as in the second case. This linear correspondence may not model the real relationship between the transistor count and the number of nets. Since the number of nets affects the number of 0-1 ILPs solved, this factor directly influences the run time.

5.8 Conclusion

In this chapter, a new global router for SOG and custom logic arrays using Zero-one Integer Linear Programming (0-1 ILP) techniques has been illustrated. The use of 0-1 ILP formulation allows *over-the-cell* channelless SOG routing without any net ordering. In the SOG router, a global routing information extraction step to capture relevant routing information and filter out extraneous details precedes the actual global routing. The nets are then decomposed into different kinds of path segments—vertical, horizontal and unrestricted—and handled separately in three routing phases. In each phase, possible path segments are identified and represented as 0-1 variables, and 0-1 ILPs are formulated to select the optimal paths that maximize the chosen benefit function. The constraints in these 0-1 ILPs are imposed by capturing important features such as the availability of routing tracks and vias over the "sea" of transistors, as well as general routing issues such as the avoidance of loops.

It has been demonstrated through experimental results of up to 108,000-transistor circuits, that the proposed method, unlike previous methods, can handle very large SOG routing problems with superior packing density and computational efficiency. It has also been demonstrated that the global router can be applied effectively to obtain compact layouts of custom logic arrays as well. The com-

5.8. CONCLUSION

putational time for various test cases was shown to be linear with the size of the circuit to be routed. The greater packing density of these circuits was the result of the ability to globally route all of the nets within the given layout area defined by the placement results of the TimberWolf program. Even large problems can be handled either directly by using larger GRC sizes, or indirectly, by partitioning the circuits into sub-blocks, and then solving each sub-block using smaller GRCs. Parallel processing can be used to exploit the inherent parallelism in the routing scheme and to speed up the runtimes.

Chapter 6

Timing-driven CMOS Layout Synthesis

6.1 Introduction

For high-performance custom VLSI chips, the layout of integrated circuits has often relied on the expertise of manual layout artists. The process of creating such manual layouts is time-consuming, tedious, and error-prone. As the size and complexity of VLSI circuits increase, the time required to create the layout, verify its correctness, and ensure that the timing specifications are met, increases drastically. At the same time, the available design cycle time has remained constant or even decreased. As a result, a strong need exists for intelligent tools to create correct layouts for various designs. Ideally, these tools should be able to generate layouts that are more compact, or at least as compact as those produced manually with a shorter turnaround time. In addition, the layout of circuits should meet all of the timing requirements specified by the designer.

In this chapter, we first describe a design methodology for standard cell design, reported in [Kan81]. The process of designing a single CMOS standard cell, which is used as a building block for a standard-cell based design, is presented. This process is used to design a set of standard cells that are placed in a library, from where they can be invoked by a designer. Next, we describe M^3 [Kan87], a metal-metal matrix layout style for automated synthesis of two-level metal CMOS circuits, which can also can be used as a platform for automated layout synthesis by a CAD tool.

Finally, we present iCGEN [Tha92], a new three-level CMOS integrated circuit layout generator, which employs the optimization algorithms discussed in earlier chapters. The layout format is based on a channelless custom logic array platform. Three layers of metallic interconnects are used to perform the routing, and a routing discipline is developed. Also, a method for mapping the triple metallic interconnects to double-layer metallic interconnects is outlined, so that the system can also be applied to two-level metal technologies. The circuit design is described to the layout generator in terms of the logic cells, a logic-level netlist, the transistor level netlist within each cell, and a set of timing specifications. A suite of program modules is executed to determine the optimal row height, the number of rows, the placement of logic cells relative to each other, the global routing of intercell signals, the actual layout of each cell and the entire design, and the optimal set of transistor sizes. These modules are iteratively executed until a compact layout, which meets all timing requirements, is generated. The user can choose to run a geometric compaction program to further reduce the layout area.

6.2 A Methodology for Designing CMOS Standard Cells

6.2.1 Introduction

MOS standard cells are often used as building blocks of large-scale random logic circuits to achieve a fast turnaround time. Although it is known to have some chip area penalty, the standard cell-based design approach allows significant savings in design efforts and time. With the availability of a standard cell library that provides information on propagation delay time and the layout of each standard cell, logic designers can provide reasonable timing specifications by estimating delay times in critical paths. The circuit designer's task then involves placing and interconnecting standard cells, subject to various timing specifications.

Although each standard cell can be designed independently, based on its design objectives, such a set of standard cells is not always optimal since each standard cell is optimized locally, with only a rough estimate of the global picture, with respect to interconnect and circuit structures.

In standard cell design, some regularity in layout style has to be maintained; otherwise, a significant amount of chip area is often sacrificed so as not to violate design rules such as the minimum separation between diffusion regions corresponding to sources and drains of p- or n- type transistors. Moreover, a uniform height is used for all standard cells, which allows easy routing of power bus lines internal to standard cells, and helps to fully utilize the routing domain. A typical standard cell layout for a two-input NAND gate is illustrated in Figure 6.1. To minimize parasitic capacitances, the horizontal pitch between polysilicon gates is kept at the minimum value allowed by the design rules. Therefore, it is clear that the major design parameters in CMOS standard cells with a uniform height are the channel widths in p- and n-type transistors, denoted W_p and W_n respectively. The separation between the p-channel and n-channel transistors, as illustrated in Figure 6.1, is denoted by S.

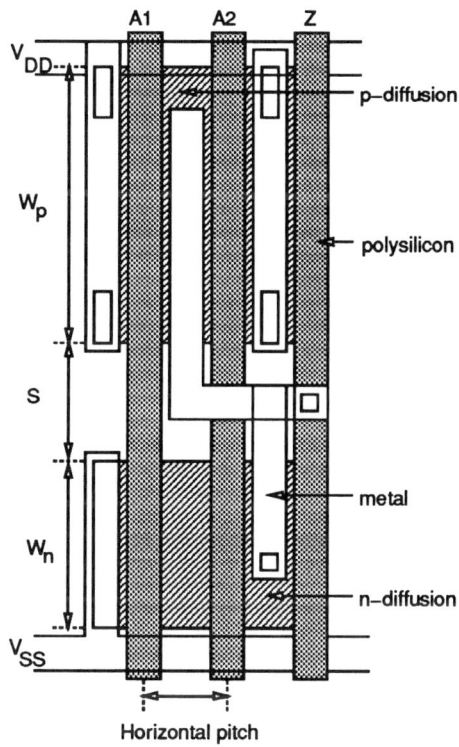

Figure 6.1: A typical layout of CMOS standard cell NAND2.

6.2.2 Design Objective

An often-discussed performance measure of integrated circuits is the product of power dissipation and propagation delay. This product is important since it represents the dissipated energy per switching operation. An equally important measure is the chip area, since the production yield of chips decreases as the chip area increases. In addition, the chip area is closely related to speed since, for a given technology, larger chips usually contain longer signal paths.

In CMOS circuits, power dissipation is small, and so far has not been a limiting factor. Therefore, for CMOS standard cell design, only propagation delay and chip area are considered.

For a given technology, the design goal is to maximize the technology power by increasing on-chip functional events per unit time. Murphy [Mur64] introduced a measure called *technology power* (TP) of the form

$$\text{TP} \triangleq \frac{\text{number of logical nodes per chip}}{\text{loaded gate delay}}. \qquad (6.1)$$

This measure can be conveniently used to evaluate performance advantages of bipolar, NMOS, CMOS, or other technologies. For instance, even though bipolar technology may yield the minimum gate delay, the number of logical nodes that can be put onto a single chip is severely limited by power dissipation. Therefore, this technology does not always give the best TP. In standard cell-based CMOS circuits, TP can be maximized by minimizing the product of active chip area and propagation delay, since the number of logical nodes on a chip bears an inverse relation to the chip area occupied by standard cells. To ensure noise immunity of circuits at 5 V and below, noise margins are required to be at least 25% of V_{DD}.

6.2.3 Basic Assumptions

For standard cell design, the following basic assumptions are made:

1) Circuit performance of a standard cell-based CMOS chip may be represented by a set of inverter (INR), two- and three-input NAND and NOR gates (henceforth denoted by NAND2, NOR2, NAND3 and NOR3).

2) On the average, the fanout of each gate is three, and the routing distance for each fanout is approximately 70 horizontal pitches.

3) The worst-case ambient temperature is 85° C, and power supply voltage V_{DD} is 5 V $\pm$ 5%.

4) For CMOS chips, power dissipation is not a limiting factor.

6.2.4 Propagation Delays in CMOS Gates

The propagation delay of an individual gate is determined by averaging 50%-point delay time in the n-channel (τ_n) and the p-channel (τ_p) case, as shown in Figure 6.2. For gates with more than one input, it will be assumed that only one input changes its state at any given time, and all other inputs will be set at V_{DD} for NAND gates, or $V_{SS} = 0$ for NOR gates.

Propagation delay times, τ_n and τ_p, decrease as the current-driving capability of the driver (proportional to gate channel widths

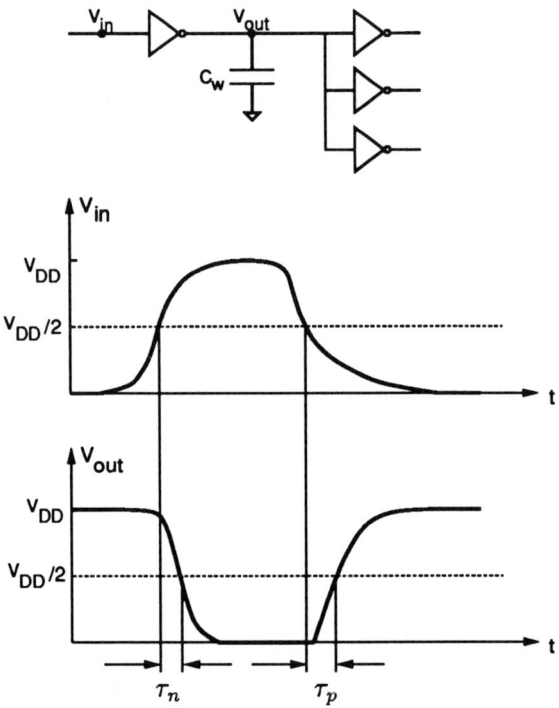

Figure 6.2: 50% point propagation delays, τ_n and τ_p, in a CMOS inverter with a fanout of 3.

W_n and W_p) increases, whereas they increase with capacitive loading, which consists of the drain capacitance in the driver (C_0), the interconnection wire capacitance (C_w), and the total input capacitance of fanout gates (C_i). Propagation delay times also depend on input waveforms. Usually, input waveforms with shorter rise and fall times yield less propagation delay. Therefore, the functional relationships between propagation delays and important circuit parameters can be described by

$$\tau_n = f(W_n, C_0 + C_w + C_i; V_{in}(.))$$
$$\tau_p = g(W_p, C_0 + C_w + C_i; V_{in}(.)) \quad (6.2)$$

where $V_{in}(.)$ denotes the input waveform, and C_0, C_w, and C_i are implicit functions of W_n and W_p.

Since propagation delays in multiinput gates are measured with all inputs except one set at either V_{DD} or zero, multiinput gates can be considered as "modified" inverters. Figure 6.3 shows equivalent circuit representations of NAND3 and NOR3 with their two inputs set at V_{DD} and zero, respectively. Basically, nonlinear RC circuits in the n-channel of NAND3 and the p-channel of NOR3 act as delay elements and increase both τ_n and τ_p.

The effect of W_n and W_p on τ_n and τ_p can best be illustrated through the INR circuit shown in Figure 6.4. Assuming, for simplicity, that the input voltage, $V_{in}(t)$, has an ideal pulse waveform, an analytic expression for τ_n can be derived from the following state equations:

NFET in saturation $(V_G - v_{out} \leq V_{th(n)})$:

$$C_L \frac{dv_{out}}{dt} = -\beta_n (V_G - V_{th(n)})^2. \quad (6.3)$$

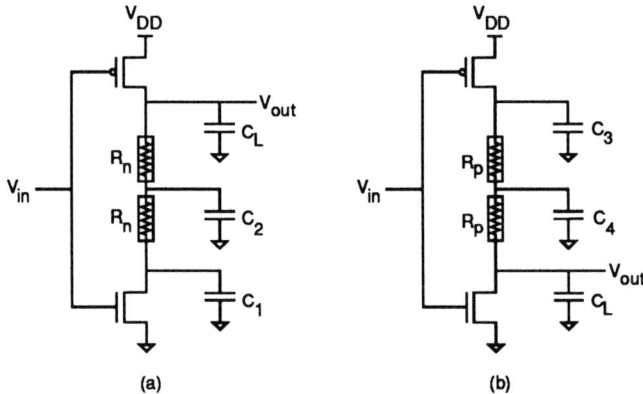

Figure 6.3: Equivalent circuits for (a) NAND3 and (b) NOR3 with two inputs set at V_{DD} and 0, respectively.

NFET in linear region $(V_G - v_{out} > V_{th(n)})$:

$$C_L \frac{dv_{out}}{dt} = -\beta_n i(2(V_G - V_{th(n)})v_{out} - v_{out}^2) \qquad (6.4)$$

where

$$\beta_n = \frac{\mu_n W_n C_{ox}}{2L_n}$$

$$C_{ox} = \frac{K_{ox}\epsilon_0}{t_{ox}}$$

The time interval for the NFET to remain in saturation, denoted by t_s, can be obtained from Eq. (6.3):

$$C_L \int_{V_{DD}}^{V_{DD}-V_{th(n)}} dv_{out} = -\beta_n(V_{DD} - V_{th(n)})^2 \int_0^{t_s} dt. \qquad (6.5)$$

CMOS STANDARD CELL DESIGN 201

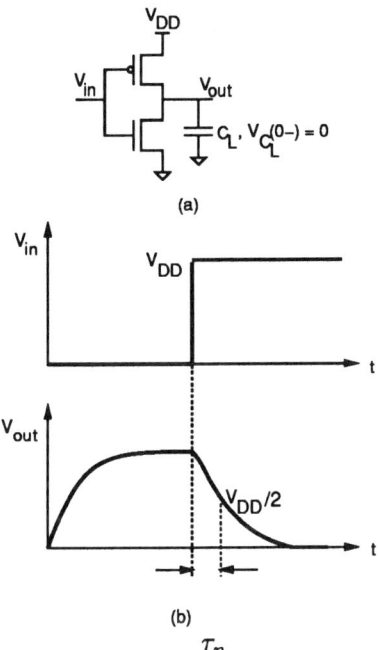

Figure 6.4: (a) CMOS inverter circuit. (b) Its input-output signal pair.

Solving Eq. (6.5) for t_s yields

$$t_s = \frac{V_th(n)}{\beta_n(V_{DD} - V_{th(n)})^2} C_L. \tag{6.6}$$

The time interval for the NFET to stay in the linear region until it reaches one half of V_{DD} is obtained from Eq. (6.4):

$$C_L \int_{V_{DD}-V_{th(n)}}^{0.5V_{DD}} \frac{dv_{out}}{2(V_{DD} - V_{th(n)})v_{out} - v_{out}^2} = -\beta_n \int_0^{t_s} dt. \tag{6.7}$$

Therefore, the propagation delay in the n-channel transistor, τ_n, has the form

$$\tau_n = \frac{C_L}{\frac{\mu_n}{2}\frac{W_n}{L_n}\frac{K_{ox}\epsilon_0}{t_{ox}}} \left[\frac{V_t h(n)}{(V_{DD} - V_{th(n)})^2} + \frac{1}{2(V_{DD} - V_{th(n)})} \right.$$
$$\left. \ln\left(\frac{1.5V_{DD} - 2V_{th(n)}}{0.5V_{DD}}\right) \right]. \quad (6.8)$$

Similarly, τ_p can be expressed by Eq. (6.8), with subscript n replaced by p.

For m-input NAND and NOR gates, propagation delays in the p-channel and the n-channel can be approximated [MC80] by

$$\tau_p(\text{NAND}m) = \tau_p$$
$$\tau_p(\text{NOR}m) = m\tau_p$$
$$\tau_n(\text{NAND}m) = m\tau_n$$
$$\tau_n(\text{NOR}m) = \tau_n.$$

Therefore, if $\tau_n^{(j)}$ and $\tau_p^{(j)}$ denote the propagation delays in the jth standard cell, among the five as ordered in assumption 1, the mean delay time in an "average" standard cell can be expressed by

$$\bar{\tau} = \frac{1}{N}\sum_{j=1}^{5}\frac{1}{2}(\tau_n^{(j)} + \tau_p^{(j)})$$
$$= \frac{1}{5} \cdot \frac{1}{2}(\tau_p + \tau_p + 2\tau_p + \tau_p + 3\tau_p + \tau_n + 2\tau_n + \tau_n + 3\tau_n + \tau_n)$$
$$= \left(K_n \frac{1}{\mu_n W_n} + K_p \frac{1}{\mu_p W_p}\right) C_L \quad (6.9)$$

where

$$K_z = \frac{1.6 L_z t_{ox}}{K_{ox}\epsilon_0}\left[\frac{V_th(z)}{(V_{DD}-V_{th(z)})^2} + \frac{1}{2(V_{DD}-V_{th(z)})} \cdot \ln\left(\frac{1.5V_{DD}-2V_{th(z)}}{0.5V_{DD}}\right)\right].$$

for $z = n$ or p. The load capacitance C_L can be expressed by

$$\begin{aligned}C_L &= C_0 + C_w + C_i \\ &= (C_{A(p)}W_p + C_{A(n)}W_n)D_d + 2(W_p+D_d)C_{p(p)} + \\ &\quad 2(W_n+D_d)C_{p(n)} + C_w(W_p,W_n) + \\ &\quad \text{fo}(W_p L_p + W_n L_n)C_{ox}.\end{aligned} \qquad (6.10)$$

For simplicity, it is assumed that the interconnection wire capacitance, C_w, remains constant as W_p and W_n are varied. In reality, any change in C_w due to changes in W_p or W_n is small; hence, this assumption is reasonable.

As mentioned earlier, the horizontal pitch between polysilicon gates is always kept at minimum so as to reduce parasitic capacitances. Therefore, the horizontal dimension that each logic gate takes, in terms of number of pitches, is fixed. For instance, INR takes only two pitches, while NAND2 and NOR2 take three pitches, and NAND3 and NOR3 take four pitches. For this reason, the effective chip area for an "average" standard cell can be represented

by
$$A = W_p + W_n + S \tag{6.11}$$

where S denotes the spacing between p-diffusion and n-diffusion regions specified by design rules. From Eqs. (6.9) to (6.11), the product of propagation delay and chip area, henceforth called the objective function, can be written as

$$H(W_p, W_n) \triangleq A \cdot \bar{\tau} = (W_p + W_n + S)\left(K_n \frac{1}{\mu_n W_n} + K_p \frac{1}{\mu_p W_p}\right)$$
$$\cdot (\alpha_p W_p + \alpha_n W_n + \alpha_0) \tag{6.12}$$

where

$$\alpha_z = C_{A(z)} D_d + 2C_{p(z)} + \text{fo} \cdot C_{ox} \cdot L_z, \quad \text{for } z = p \text{ or } n$$
$$\alpha_0 = 2D_d(C_{p(p)} + C_{p(n)}) + C_w. \tag{6.13}$$

It can be shown from Eq. (6.12) that the objective function $H(W_p, W_n)$ is a posynomial in W_p and W_n. A simple transformation (see Section 2.6) can be used to show that this function is equivalent to a convex function.

6.2.5 Consideration of Noise Margins

The noise margin of a logic gate is defined as the input voltage differences between the dc operating points and their nearest unity gain points. Without sufficient noise margins, cross-couplings between adjacent metal or poly lines on the chip, or from the external environment, may cause erroneous outputs. As design rules shrink, such cross-couplings tend to increase, and the consideration of noise margins becomes an important factor [Mur64].

For a typical INR gate, its dc transfer characteristics can be calculated using V-I characteristic equations of p-channel and n-channel transistors. In particular, the input gate voltage at which both p-channel and n-channel transistors operate in saturation can be expressed by [TS77]

$$V_{i(sat)} = \frac{\sqrt{\beta_R}[V_{DD} - V_{th(p)}] + V_{th(n)}}{1 + \sqrt{\beta_R}} \qquad (6.14)$$

where

$$\beta_R = \frac{\mu_p W_p}{\mu_n W_n}. \qquad (6.15)$$

Eq. (6.14) suggests that when $V_{th(n)} = V_{th(p)}$, the midtransition point can be centered at $V_{DD}/2$ to allow larger noise margins by setting

$$\frac{W_p}{W_n} = \frac{\mu_n}{\mu_p}. \qquad (6.16)$$

In other words, for maximum noise margins, W_p should be about three times W_n, since the effective channel mobility carriers in an n-channel is about thrice that in a p-channel.

6.2.6 Worst-case Simulations and Standard Cell Design

It is well known that transistor characteristics often depart from their nominal values due to processing variations. To check whether the design would meet the performance specifications, SPICE simulations are done under worst-case conditions to calculate propagation delays and dc transfer characteristics.

Worst-case device parameters that yield a low current-driving capability and, therefore, longer propagation delays, are fed into SPICE through a so-called "worst-case technology file". Also, the power supply voltage V_{DD} is set at 4.75 V, 5% below its nominal value, and the device junction temperature is set at 95° C to account for on-chip power dissipation, although the ambient temperature is assumed to be 85° C. With the increase in chip complexity, more conservative specifications, such as $V_{DD} = 4.5$ V and $T_j = 105°C$, would have to be used. Two extreme cases are considered to ensure sufficient noise margins. One extreme case is simulated with high-current p-channel and low-current n-channel device parameters and the other extreme case with low-current p-channel and high-current n-channel device parameters.

Figure 6.5 illustrates the simulation approach that is used to achieve the design goal, for the case of 3.5 μm design rules, and

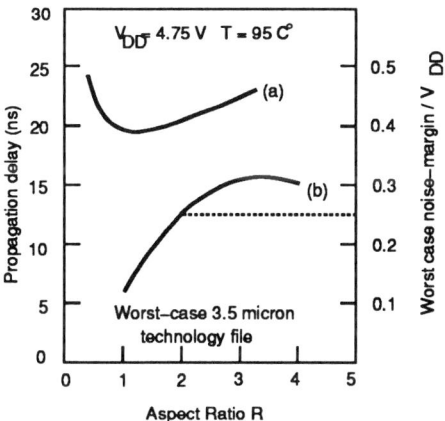

Figure 6.5: (a) The propagation delay (ns) of an average logic gate. (b) The worst-case noise margin/V_{DD} vs. the aspect ratio $R \triangleq W_p/W_n$.

worst-case device parameters.

Simulation results shown in Figure 6.5 are obtained with $W \triangleq W_p + W_n = 52\mu m$. Curve (a) represents the average of worst-case propagation delays in INR, NAND2, NOR2, NAND3 and NOR3 for different aspect ratios between W_p and W_n, while curve (b) shows worst-case noise margins. Curve (b) suggests that the aspect ratio $R \triangleq W_p/W_n$ should not be less than 2, in order to ensure the worst-case noise margin to be 25% of V_{DD}. Although the maximum noise margin can be achieved by setting $R = 3$, such a choice is not desirable as the minimum propagation delay occurs for R between 1 and

2. Since the design goal is to minimize the delay-area product, and thereby to maximize the technology power, $R = 2$ is chosen to avoid degrading the switching speeds of CMOS standard cells. Figures 6.6 and 6.7 show dc transfer characteristic curves with $W_p/W_n = 1$ and $W_p/W_n = 2$, respectively. At $V_{DD} = 2$ V, the noise margin can be as low as 0.3 V, with $W_p/W_n = 2$. It can be observed that dc characteristic curves shift to the right as the aspect ratio R increases, and achieves the maximum noise margin at $R = 3$.

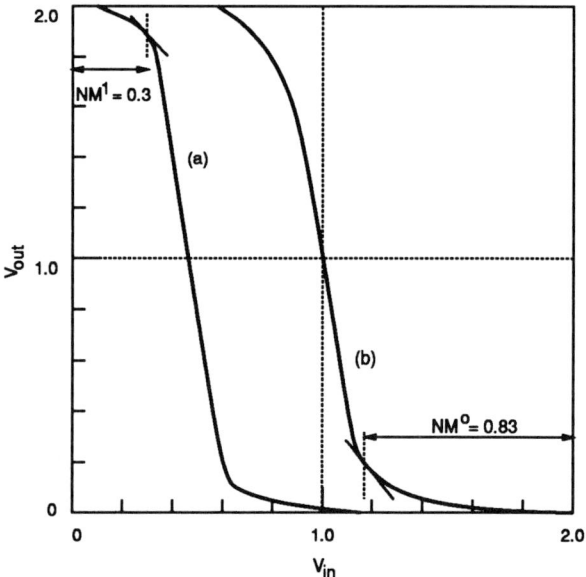

Figure 6.6: dc transfer characteristics of INR with $W_p/W_n = 26/26$. (a) High-current n-channel and low-current p-channel parameters. (b) Low-current n-channel and high-current p-channel parameters.

Having determined that the aspect ratio R should be 2, the chan-

CMOS STANDARD CELL DESIGN

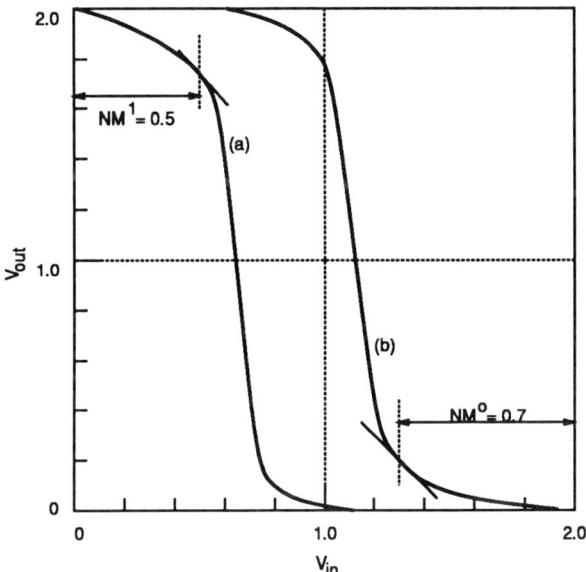

Figure 6.7: dc transfer characteristics of INR with $W_p/W_n = 35/17$. (a) High-current n-channel and low-current p-channel parameters. (b) Low-current n-channel and high-current p-channel parameters.

nel widths W_p and W_n that minimize the delay-area product must be determined. As shown in Figure 6.8, the propagation delay decreases with increasing channel widths at the cost of chip area. Note that with a chosen aspect ratio $R = 2$, the problem becomes a one-dimensional search. In fact, Eq. (6.12) can be rewritten in terms of the total channel width W as

$$H(W) \triangleq A \cdot \bar{\tau} = (W + S) \cdot \frac{\alpha_{-1}}{W}(\alpha_1 W + \alpha_0) \qquad (6.17)$$

where

$$\alpha_{-1} = \frac{K_n}{\mu_n}(R+1) + \frac{K_p}{\mu_p}\frac{R+1}{R}$$

$$\alpha_1 = \alpha_p \frac{R}{R+1} + \frac{\alpha_n}{R+1}.$$

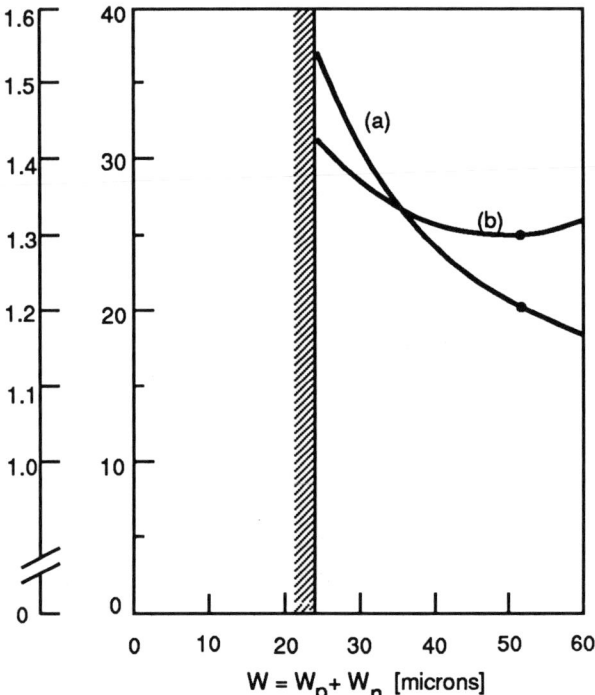

Figure 6.8: (a) The average-gate propagation delay (ns). (b) The delay time-chip area product (10^{-12}s.m^2) vs. the total channel width $W = W_n + W_p$.

In Eq. (6.18), the chip area A is an affine function of W, and the mean propagation delay is composed of a fixed term, and a term

CMOS STANDARD CELL DESIGN

that is inversely proportional to W. Therefore, for small values of W, the speed improvement is quite noticeable as W increases, while the increase in chip area is small. Physically, it is indeed necessary to increase transistor sizes beyond the minimum size to overcome capacitive loadings from interconnection wires. Without routing capacitances, the best design would be to use minimum size transistors. On the other extreme, if the transistor sizes became excessively large, the gate loading from fanout gates would be dominant over routing capacitances and the propagation delay would remain almost constant, leaving the increase in chip area unjustified. Indeed, SPICE simulation supports the model in Eq. (6.18), and finds the minimum point on curve (b) at $W = 52$ μm. The shaded region in Figure 6.8 is not accessible due to design rules. This design yields the sizes of p-channel and n-channel transistors to be $W_p = 35$ μm, and $W_n = 17$ μm. It also allows up to six metal wiring tracks internal to the standard cells. Figure 6.9 shows the worst-case propagation delays in standard cell circuits for different values of the capacitive load C_L.

6.2.7 Conclusion

A design of CMOS standard cells for random logic circuits has been considered. The design objective is to minimize the area-delay product, allowing sufficient noise margins under worst-case conditions.

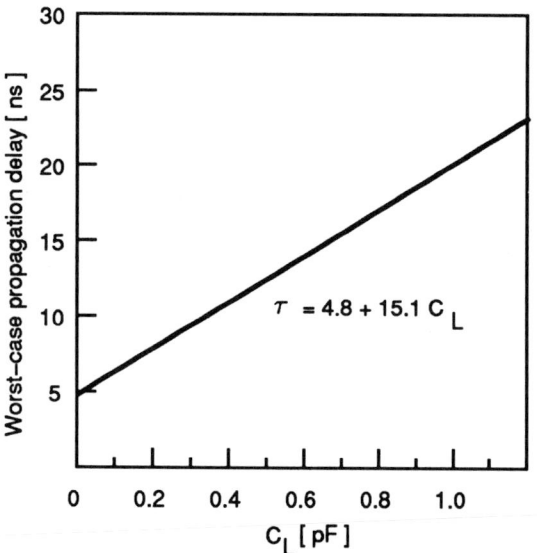

Figure 6.9: The worst-case propagation delay, τ, of an "average" standard cell gate versus capacitive loading C_L.

This design objective is in line with maximizing technology power. The designable parameters in CMOS standard cells with a uniform height are identified as the n-channel and p-channel transistor widths, W_p and W_n, respectively. An analytic expression for the design objective function in terms of the designable parameters, W_p and W_n, is derived. It is shown that the objective function is equivalent to a convex function and, hence, has a unique minimum. A physical interpretation of the model is given in detail.

The worst-case propagation delays and noise margins are calculated using SPICE, for INR, NAND2, NAND3, NOR2 and NOR3 gates with an average fanout of three, under the assumption that the routing distance for each fanout is 70 horizontal pitches. With the power supply voltage, V_{DD} taken as 5 V ± 5%, and the ambient temperature taken to be 85° C, standard cells are designed to have $W_p = 35$ μm and $W_n = 17$ μm, for 3.5 μm design rules. This allows the worst-case noise margin to be at least 25% of V_{DD}.

Although the material in this section deals with typical logic gates, other standard cells can be conveniently designed with the same design parameters, except special standard cells used for input-output (I/O) pads. Usually such I/O standard cells have to be very large to provide a sufficient current-driving capability. Buffers of different sizes can be designed by properly connecting transistors of uniform heights in parallel. If there were any logical node requiring more than average fanouts, such a node could be handled by using a high-power version of the driving standard cell, which could also be implemented with a uniform height.

6.3 The Metal-Metal Matrix (M^3) Layout Style for Two-level Technologies

Since interconnect delays are a significant part of the circuit delay, it is important to develop a layout discipline that ensures that higher-resistance polysilicon lines are used only for short-distance connec-

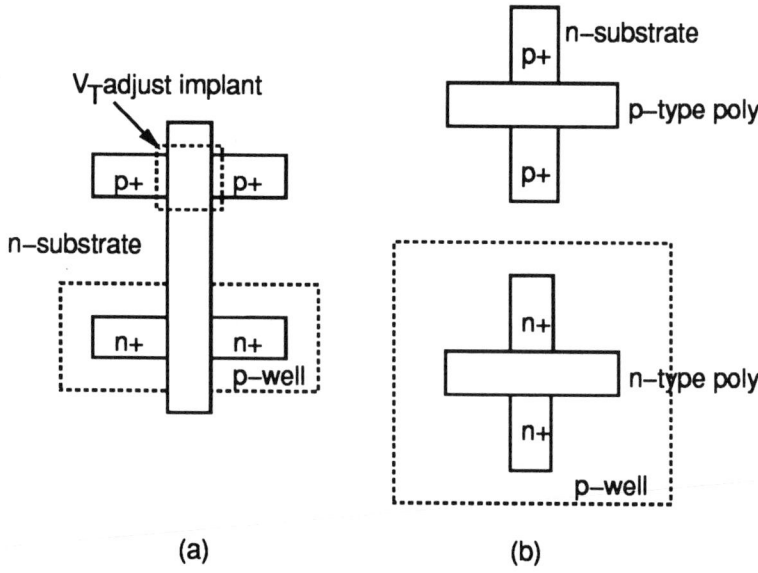

Figure 6.10: Two layout structures for the inverter circuit.

tions, and that all long-distance connections are made with metal lines, which have relatively lower resistances. A methodology for two-level metal technology, called metal-metal matrix (M^3) layout, is presented in [Kan87]. The main features of this approach are:

(1) A minimal amount of poly is used, to suppress parasitic RC delays.

(2) Vertical stacking of p-type and n-type transistors with contiguous poly, shown in Figure 6.10(a), is avoided, and the configuration in Figure 6.10(b) is used instead. The reason for this is as follows. The threshold voltage of a transistor can be controlled by an implantation step that dopes the polysilicon gate

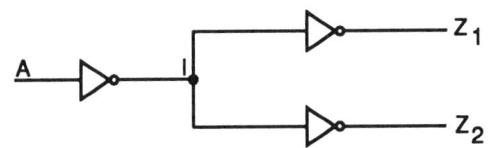

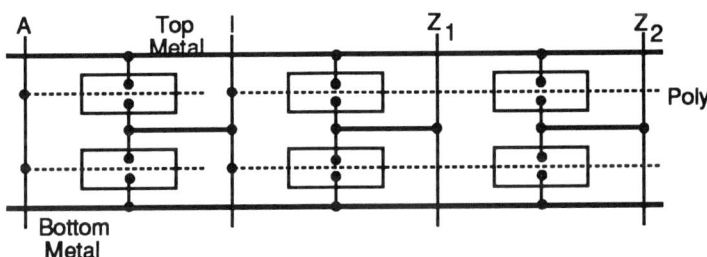

Figure 6.11: Layout example using M^3.

of a transistor. In future generation CMOS technologies with lower power supply voltages, it is likely that better threshold voltage control would be required. This would entail doping the poly gate material separately for n-type and p-type transistors, which would require transistors to be laid out as shown in Figure 6.10(b).

(3) To accommodate multiple fanouts, common poly lines among the same type of transistors are allowed.

(4) Metal runners are used for all other interconnections.

(5) The terminals of all modules or standard cells are made of metal; the use of poly for such terminals is strictly avoided.

A simple layout example is shown in Figure 6.11. Note that this layout could be optimized further by intelligent use of CAD tools. A slight variant of M^3 has been used as the layout platform for the timing-driven CMOS module synthesis system, PERFLEX [KOI92], which integrates transistor sizing, transistor reordering, and interconnect parasitic reduction, and generates an automated layout that satisfies the timing requirements.

6.4 iCGEN : A CMOS Layout Synthesis System for Three-level Metal Technology

6.4.1 Outline of iCGEN

Figure 6.12 provides an overview of the entire layout process which consists of initial transistor sizing, determination of an optimal row height and the number of rows, placement of logic cells, global routing of signals, layout generation, and transistor sizing. The organization of the iCGEN system is modular, and thus new or improved versions of modules can be readily exchanged with the existing ones, as long as their basic functionality remains unchanged.

The circuit is described in terms of a logic-level netlist, the interconnection of transistors within each logic cell, and the timing specifications. For the first pass through the layout process, the

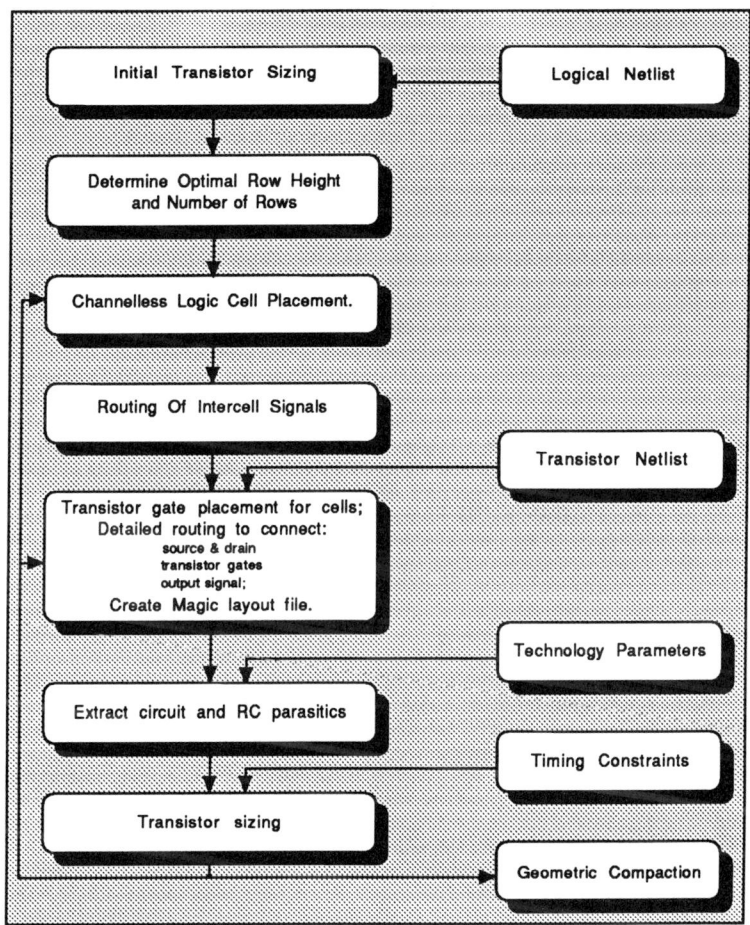

Figure 6.12: An overview of the layout system.

initial transistor sizes can either be specified by the designer or calculated by the transistor sizing module. Since the system iterates the layout process until a stable set of transistor sizes is reached, it is desirable that the initial sizes should be close to the final sizes, in order to reduce the number of iterations required. Hence, it is advisable to use the transistor sizing program, even before layout, with a rough estimate of the RC parasitics. Once the initial transistor sizes are determined, the next step is to determine the optimal row height and the number of rows. Since the layout style is based on a regular platform shown in Figure 6.13, the exact size of the initial layout can be computed for the initial set of transistor sizes. A table which shows the area and aspect ratio of the design for various row heights and row counts can then be constructed. From such a table, the designer or the program, by default, can choose the appropriate number of rows and the row height, which minimize the total area for a given aspect ratio. This information is then used to determine the size of each logic cell, taking into account the need for folding large transistors.

After the cell sizes are determined, logic cells are placed into rows with a goal of minimizing the total length of the interconnection wires. The main purpose of this phase is to determine the relative locations of the logic cells, rather than their absolute locations.

iCGEN : A CMOS LAYOUT SYNTHESIS SYSTEM

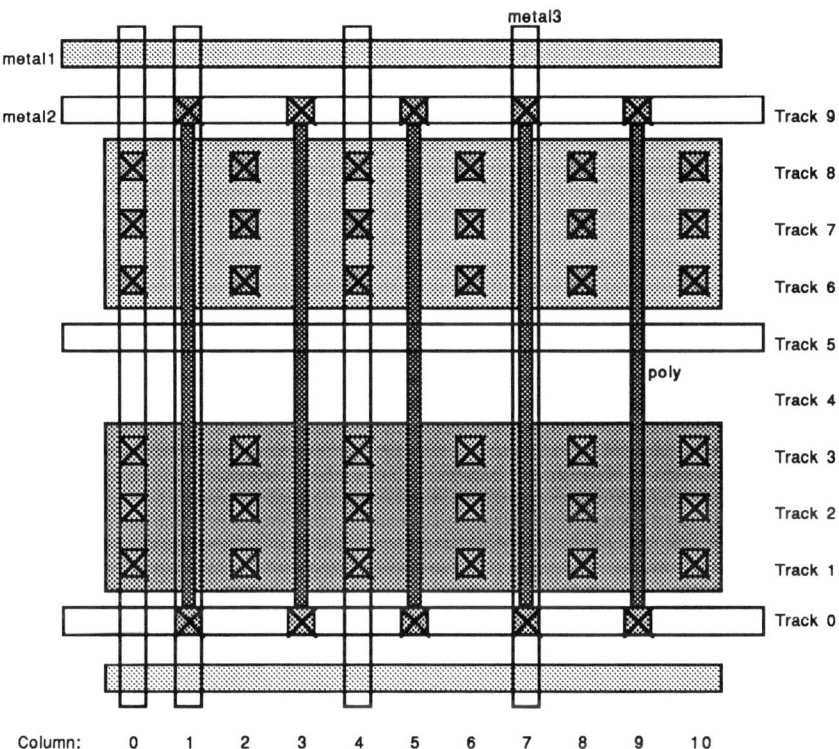

Figure 6.13: Base cell of the logic platform.

Standard placement programs can be adopted in this phase. This information, along with the logic-level netlist, is then used to generate the information needed in the global routing phase.

After the cell placement, a global routing step is performed to determine the coarse placement of signal nets over the layout platform. Initially, an array of rectangular global routing regions is superimposed over the area where the logic cells have been placed. The terminals of signals connecting cells are placed into the appropriate routing regions. The global routing program then uses zero-one linear integer programming techniques [HS85b], described in Chapter 5, to determine the set of rectangular routing regions that each net should pass over to connect all of the terminals for that net. By adjusting the size of the routing regions and partitioning the circuit, very large circuits can be routed efficiently.

The layout of each logic cell is then generated in accordance with the data format of the layout tool *MAGIC*. This module first assigns tracks to all of the signals which were globally routed over the cells in the previous step. The placement of the signals into the tracks then serves to create the boundary conditions for the various cells in the circuit. The module then proceeds to generate the layout of each logic cell, taking into account the need for transistor folding and the boundary conditions. A transistor is folded if its size is too large to fit into a contiguous diffusion strip. For transistors smaller than the nominal width of a diffusion strip, the width of the diffusion of each transistor is trimmed back to the correct size. The first step in generating the layout for a cell consists of finding an optimal ordering of transistor gates such that the minimum number of diffusion breaks is needed. If multiple orderings lead to the same minimum number of diffusion breaks, the ordering which requires the minimum amount of routing is chosen. Once an ordering is found, the system begins to generate the layout of all of the cells in the design in real time. After all of the cells are created and connected, the entire circuit is extracted from the layout for exact transistor sizing to meet the user-specified delay specifications.

The next step is to determine the size of all of the transistors, such that the total sum of the transistor sizes is minimized, while meeting all timing requirements. The program uses iCONTRAST, discussed in Chapter 4, to find the optimal sizes for the transistors. Once the new transistor sizes are found, the layout of the circuit has to be updated accordingly. Depending on how extensively the transistor sizes change, the process may return to the cell placement phase or the cell layout generation phase. If the changes are large enough to cause changes in the size of the logic cells, due to the need to fold or unfold certain transistors, the process will return to the cell placement stage. However, with an initial transistor sizing based on the given netlist, such backtracking is rare. If the changes are minor and can be handled by adjusting the sizes of the existing diffusions, only the cells requiring changes will be modified locally. This process is repeated until it converges to a stable set of transistor sizes, which usually takes only a few iterations. This process always converges as long as the boundary conditions do not cause routing problems due to insufficient tracks. If desired, a post-processing geometric compaction program can be used to produce the final layout masks.

6.4.2 Layout Platform

The quality of any layout system is dependent on how well it assigns and utilizes the routing resources available to it. A good routing scheme will assign the various metallic layers to specific uses and optimal directionality. Once a good routing scheme is established, the process of creating the detailed layout is greatly simplified and should result in a compact layout using a minimal amount of computation. The layout system is geared for three layers of metallic interconnect, but a mapping to two layers can be carried out for the double-metal layer technology.

The platform used in generating the layout of the design is shown in Figure 6.13. This platform is similar to the sea-of-gates array structure proposed by Duchene and Declercq [DD89], but produces much more compact layouts. The reason for this improvement in layout area stems from the fact that the new layout system assumes full control over all of the geometries for all of the layers in the process, while the sea-of-gates approach has control over only the metal layers to form the interconnect. The layout platform style was chosen because of the regularity of its structure and the compactness of the layouts created.

For each cell to be created, the locations of all of the input and output signals are determined by the track assignment step, which follows the global routing step. All of the globally routed signals will be carried by a span of horizontal metal2 wire crossing over the entire cell. The detailed routing connects these globally routed signals to the polysilicon input gates and the output drain diffusions. In addition, the detailed routing realizes the correct connectivity of diffusion nodes within each cell.

Metal1 Usage

The first layer of metal, metal1, is used for intracell routing and the distribution of power and ground to the cells. The use of metal1 within each cell is to form connections between the sources and drains of transistors, to connect the output diffusion drains of the PMOS and the NMOS transistors together to the globally routed signal, and to help connect the polysilicon gates to the globally routed input signals.

The primary use of metal1 is to form horizontal connections between the sources and drains of transistors. For this purpose, the strips of metal1 are restricted to lie within the n-diffusion, in horizontal tracks 1-3 in Figure 6.13, and within the p-diffusion, in horizontal tracks 6-8.

Metal1 is also used to form connections between the polysilicon gate contacts and the globally routed input signals. For this purpose, a series of short horizontal strips are formed in tracks 0 and 9. A via from metal3, which will carry the input signal, to metal1 is created in each even-numbered column, i, prior to the polysilicon gate. Then a piece of metal1 is used to connect the via to the polysilicon contact in each odd-numbered column, $i+1$.

In addition, metal1 is used to connect the drain regions of the NMOS and the PMOS transistors forming the output node of the cell to the globally routed signal. To create this interconnection, both vertical and horizontal strips of metal1 are used. First, a vertical strip spanning across horizontal tracks 3-6 in Figure 6.13 is created to connect the two diffusion nodes together. It may be noted that this area between the two diffusion nodes will be devoid of metal1 strips and hence free of conflicts. Next, a horizontal strip is formed from the vertical strip to the last column of the cell. This connection will reside in the track immediately above the n-diffusion strip, track 4 in the figure. The last step is to form a vertical connection in the last column of the cell, to connect the horizontal strip just formed to the globally routed output signal.

Metal2 Usage

The second layer of metal, metal2, is used solely for the routing of global intercell signals. Its purpose is to form the horizontal interconnections over the cells within each row of the layout. Basically, all of the input signals and the output signal of every cell will be carried by a piece of metal2. In creating the various strips of horizontal interconnect, the use of metal2 is restricted to the area over and in between the two diffusions, i.e., tracks 1-8 in Figure 6.13. This will ensure that the vias from metal3 to metal1 in the track above the p-diffusion and the track below the n-diffusion can be created without forming shorts.

Metal3 Usage

The third layer of metal, metal3, is used to form both the global interconnections between cells and the local connections within each cell. The use of metal3 is in the vertical direction only. To ensure that both the global and local metal3 strips can be formed, the two different uses are constrained to reside in certain vertical tracks. All global interconnections are limited to lie in the vertical tracks over the polysilicon gates, and all local interconnections are limited to lie in the vertical tracks in between the polysilicon wires. In Figure 6.13, this would correspond to the global signals residing in odd-numbered columns, while the local interconnects resides in even-numbered columns. To form the global interconnections, vertical strips of metal3 are formed to connect signals spanning different rows. Metal3 is also used to form local interconnections of globally routed input signals to the gates of the transistors. For this purpose, a via from metal2 to metal3 is created in each vertical track prior to the transistor gate in the appropriate horizontal track. A piece of metal3 is then formed in the same vertical track, from the track just below the n-diffusion to the track just above the p-diffusion. This span of metal3 is then connected to a piece of metal1 which, in turn, connects to the polysilicon gate.

6.4.3 Mapping from Triple Metal Layers to Double Metal Layers

If only two layers of metallic interconnects are available for routing purposes, a mapping of a three-layer layout into a two-layer layout can be used. Since both metal1 and metal2 run mostly in the horizontal direction, they can be mapped onto the first layer of metal in the horizontal direction, with some modifications to resolve any potential conflicts. The vertical metal3 strips can then be mapped directly into the second layer of metal in the vertical direction.

6.4.4 Determination of Optimal Row Height

One advantage of the layout platform described in the previous section stems from its regularity. Using this structured layout style, the total area and the aspect ratio of a design can be computed before the entire layout is actually generated. The ability to calculate such values allows the designer to determine whether the current design can meet the constraints on the area and the aspect ratio. This ability can save designers both time and effort which would be wasted in generating layouts that do not meet the required constraints.

Before determining the total area of a design, several pieces of information are required. First, the total number of cells and the content of each cell have to be known. For each cell in the design, the size of each n-transistor, W_n, and each p-transistor, W_p, in the cell should be supplied before the total area can be determined. In addition, a range for the number of rows, *numrows*, in the final design should be supplied. The program will then calculate a range of values for the total width of both diffusions combined in each row, W_r. Given this information, the following algorithm is used to determine the final layout area of the design:

```
Set the total width of all the cells, W_t, to zero.
For each cell in the design, do {
    Set the width of this cell, W_c, to zero.
    Read in the size of each transistor in the cell and determine the
    total width of the cell (i.e., number of polysilicon lines needed.)
    For each n-p transistor pair in the cell, do {
        Determine the number of vertical polysilicon lines needed to
        lay out these two transistors.
        If W_n + W_p ≤ W_r then
            W_c = W_c + 1      /* Only one polysilicon line needed */
        Else
            W_c = W_c + ⌈(W_n+W_p)/W_r⌉.  /* Folded transistors */
    }
    W_c = W_c + 1.    /* Extra space for separation between cells */
    W_t = W_t + W_c.  /* Maintain running total of cell widths */
}
While numrows is in the range of the specified number of rows, do
    Determine the aspect ratio and total area.
```

To determine the aspect ratio and area of the layout for the varying number of rows, the width and height of the layout can be determined with the following equations:

$$X = \left\lceil \frac{W_t}{numrows} \right\rceil * 16 \qquad (6.18)$$

$$Y = numrows * (W_r + 44) \qquad (6.19)$$

$$R = \frac{X}{Y}, \qquad (6.20)$$

where X is the total width of the layout area in terms of λ,
W_t is the total width of all of the cells in terms of the number of polysilicon lines needed,
$numrows$ is the number of rows in the design,
the space between adjacent polysilicon lines is 16 λ,
Y is the total height of the layout area in terms of λ,
W_r is the width of the diffusion area in each row,
the addition vertical height needed in addition to W_r for each row is 44 λ,
and R is the aspect ratio of the design.

The algorithm given above is executed for a range of W_r bounded by the smallest n/p transistor pair and the largest n/p transistor pair, $(W_n + W_p)_{min} \leq W_r \leq (W_n + W_p)_{max}$. The reason for starting with the lower bound on $W_n + W_p$ stems from the fact that if a smaller value were chosen, all of the transistors in the design would be too large to fit into a contiguous strip of diffusion. In such cases, all of the transistors have to be folded, which would not lead to a good layout. As the value of W_r increases, the total height of the design also increases, but the number of transistors which require folding will start to decrease at certain points. With this decline in the number of folded transistors, the width of the design will also shrink until W_r equals the maximum $W_n + W_p$. At this point, no transistor folding is needed. Thus, increasing W_r beyond this point would not shrink the layout's width any further, whereas the height would increase and also the total area of the design. All of the data concerning the width, height, and aspect ratio of the layout for different W_r and *numrows* is then entered into a table. From this table, values for W_r and *numrows* can be chosen to either minimize the total layout area, or to meet a certain aspect ratio, or both.

6.4.5 Channelless Logic Cell Placement

Once the number of rows and the widths of both diffusions in each row have been selected, the exact dimensions of each logic cell can be calculated after determining which transistors need to be folded. After the dimension of each cell is known, the next phase of the layout process deals with the butting placement of logic cells with no separate channels allocated for routing. The goal of placement is to minimize the total length of wires interconnecting cells over the cell area, while maintaining a uniform width among the rows. The result of the placement phase should be an assignment of logic cells into the rows of the design and the location of each cell within its row. The cell placement program, *TimberWolfSC* version 5.4 [SSV85], is used to perform the placement of logic cells in iCGEN, but any other placement program could be used instead. Figure 6.14 shows an example of butting cell placement without channels.

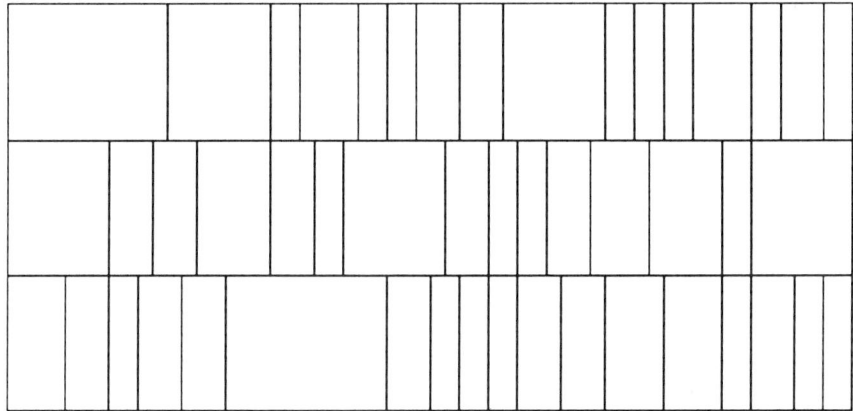

Figure 6.14: Cell placement without channels.

6.4.6 Routing of Intercell Signals

After the placement of logic cells has been completed, the next step in the layout process is to route the signals connecting the logic cells. This routing is performed in two stages. First, the global routing phase determines the coarse location of the wires over the layout platform. A program employing zero-one integer linear programming (0-1 ILP) techniques [KB80] is used to choose the best set among many alternative routing scenarios. After global routing has been performed, the track assignment phase is entered to determine the exact location of wires. The track assignment is done using a left-edge track assignment algorithm.

Before the global routing of nets is performed, an array of rectangular routing regions, called global routing cells (GRCs), is superimposed onto the cell placement. For each net, the locations of its terminals are extracted and placed into the GRCs. This extraction produces information on the location of terminals in various GRCs for a set of nets. The global routing phase then determines the placement of wires into the GRCs, such that all of the terminals of each net are connected, while not exceeding the routing resources available in each region. The global routing phase uses 0-1 ILP to formulate and solve the routing problem. The 0-1 ILP used to determine the location of the vertical metal span can be formulated as follows:

$$\text{MAXIMIZE} \sum_{\text{all } n} \sum_{\text{all } j} B_{n,j} \, P_{n,j}^{m3}$$

SUBJECT TO

Path Usage Constraints:
for each net, n,
$$\sum_{\text{all } j} P_{n,j}^{m3} \leq 1$$

South Boundary Constraints:
$\forall i, j, k$, for the (i,j,k)th region,
$$\sum_{\text{all } n_{i,j,k}} P_{n_{i,j,k},j}^{m3} \leq S_{i,j,k}^{m3}$$

Via Constraints:
$\forall i, j, k$, for the (i,j,k)th region,
$$\sum_{\text{all } n_{i,j,k}} P_{n_{i,j,k},j}^{m3} \leq V_{i,j,k}$$

AND *0-1 Variable Constraints:*
$$P_{n,j}^{m3} = 0 \text{ or } 1$$

where $P_{n,j}^{m3}$: the path of net n in column j;
$S_{i,j,k}^{m3}$: the metal-3 South Boundary Capacity of the (i,j,k)th region;
$V_{i,j,k}$: the Via Capacity of the (i,j,k)th region;
and $B_{n,j}$: the *benefit* value of partially connecting the net n using the path in column j.

The global routing performed results in structures which possess one vertical metal3 span, if the net has terminals in different rows, from which horizontal metal2 spans will emanate for each row in which the net has terminals. The above problem corresponds to the Phase One of the sea-of-gates global routing algorithm presented earlier; further details are provided in Chapter 5.

After the global routing has been performed, the exact locations of nets are determined by using a left-edge track assignment algorithm. For each row/column of regions, all of the nets with spans in that set of regions are found and sorted by the starting points of their spans. Once this sorted list has been extracted, each net in the list is assigned to the first available track. Since a successful global routing phase ensures that none of the routing resources are exceeded, the track assignment algorithm is guaranteed to work. Once the exact locations of all wire segments forming the nets are determined, the next phase of the layout process is the generation of each logic cell.

6.4.7 Logic Cell Layout Generation

After the routing of global signals has been performed, the detailed layout of each logic cell has to be created. The first step in generating the layout is to determine the optimal ordering of the transistor gates in each cell so that the width of the cell is minimized. Since the system uses the layout platform shown in Figure 6.13, the width of each cell can be minimized by chaining the sources and drains of adjacent transistors when they are electrically equivalent. If the two diffusion areas are not electrically equivalent, then a break in the diffusion is required. This break would cause the width of the cell to increase. Thus, the goal of this step is to find an ordering of the transistor gates such that a minimum number of diffusion breaks are needed. To determine such an ordering, graph models for the p-net and the n-net are used for each logic cell [UV81]. A node in the graph corresponds to an electrical node in the cell. An edge represents a transistor connecting two nodes. For large transistors requiring k folds, k parallel edges are placed in the graph model. Then, an Euler path for both graphs has to be found. If such a path exists, then the ordering of transistors, based on the order of edges in the Euler path, leads to a layout without any diffusion breaks. If no such Euler path exists, then the two graphs can be partitioned into smaller subgraphs for which Euler paths exist. With this scheme, the ordering of transistors in each subgraph does not require any diffusion breaks, but diffusion breaks are required between transistor groups of different subgraphs.

After an optimal ordering is found, detailed routing is performed. This routing consists of forming source/drain connections in the n-diffusion and p-diffusion regions, connecting the globally routed input signals to the polysilicon input gates, and connecting the two output diffusion drains to the globally routed signal.

Source/drain routing has to be performed to connect electrically equivalent, but spatially separated, diffusion areas for both the n-net and p-net. For each node that requires source/drain routing, the starting and ending point of the metal1 span can be found by scanning the sequence of nodes in the Euler path. Once all of the metal1 spans are found, they are sorted according to their starting points and assigned to tracks using the left-edge algorithm.

6.4.8 Circuit Tuning Using Convex Optimization

Given the circuit netlist and a set of user-specified delay specifications, the transistor sizing module, iCONTRAST, attempts to find the best set of transistor sizes that satisfies the specifications.

A typical digital integrated circuit consists of multiple stages of combinational logic blocks that lie between latches, clocked by system clock signals. It must be ensured that the worst-case delay of the combinational blocks is such that valid signals reach a latch before any transition in the signal clocking the latch, with allowances for set-up time requirements. For a combinational circuit, the transistor sizing problem is formulated as

$$minimize \quad Area \qquad (6.21)$$
$$subject\ to \quad Delay \leq T_{spec}$$

As discussed in Chapter 4, the active area, measured as the sum of transistor sizes, and the delay along a path of the circuit can be represented by *posynomial* functions of the sizes of transistors in the circuit. A posynomial is a function g of a positive variable $\mathbf{x} \in \mathbf{R}^n$ that has the form

$$g(\mathbf{x}) = \sum_j \gamma_j \prod_{i=1}^n x_i^{\alpha_{ij}} \qquad (6.22)$$

where the exponents $\alpha_{ij} \in \mathbf{R}$ and the coefficients $\gamma_j > 0$. Such a function can be mapped onto a convex function through an elementary variable transformation, $(x_i) = (e^{z_i})$ [Eck80].

The delay of a circuit is computed as the Elmore delay of a worst-case RC network that represents the circuit. Our delay model computes rise and fall delays separately, and is capable of handling waveforms with nonzero transition times.

The delay of a circuit, defined here to be the maximum of the delays of all paths in the circuit, is the maximum of posynomial functions, which is mapped by the above transformation onto a maximum of convex functions. The area function is also transformed into a convex function by the same mapping. Therefore, the optimization problem defined in (6.21) is mapped to a *convex programming* problem, i.e., the problem of minimizing a convex function over a convex constraint set. Due to the unimodal property of convex functions over convex sets, any local minimum of (6.21) is also a global minimum.

An efficient convex programming method [Vai89] is used for global optimization over the parameter space of all transistor sizes in a combinational subcircuit, thereby solving the problem *exactly*. The algorithm starts by bounding the convex domain by an initial polytope. By using special cutting plane techniques, the volume of this polytope is shrunk in each iteration, while ensuring that the optimal solution lies within the boundary of the polytope. The iterative procedure stops when the volume of the polytope becomes sufficiently small. This algorithm does not require enumeration of the constraints defining the feasible region, which is a major advantage over other transistor size optimization algorithms. A complete description of this algorithm is provided in Chapter 4.

After the new transistor sizes are determined, a new layout is generated if any of the transistor sizes changed. Depending on the extent of the change, the layout process returns to either the cell placement phase or the layout generation phase. The layout process then iterates until a compact layout meeting all of the timing specifications is created. This process converges in a few iterations.

6.4.9 Experimental Results

The layout synthesis program was used to benchmark a combinatorial circuit provided from industry. The circuit contained 190 unique transistors and 43 intercell signals. Using the initial transistor sizes provided for the circuit, the area of the automated layout, shown in Figure 6.15, was 928 λ x 444 λ (= 412032 λ^2). Here, λ represents the feature size of the technology. This layout was approximately 1% smaller than the manually created layout from industry, which was 811 λ x 514 λ (= 416854 λ^2), even without geometric compaction. This layout maintained the same transistor sizes as the industry layout for comparison purposes. If geometric compaction were to be performed, the reduction in area would be estimated to be around 15 to 20 percent. The transistor sizing module was then exercised to resize the transistors in the circuit under the same timing specifications. After the transistor resizing, the new layout was 512 λ x 646 λ (=331776 λ^2), which is approximately 80% of the manually created circuit. This suggests that the transistor sizing of the manually laid-out circuit is suboptimal.

The layout of another industrial circuit, shown in Figure 6.16, was also generated. The circuit was composed of complex AOI and OAI gates and contained 204 transistors. The total area of the circuit layout was 600 λ x 432 λ (=259200 λ^2), which is significantly more compact than the gate-matrix layout [LL80] from industry.

A set of tests was run to determine the area versus speed characteristic of the layout system. Multiple layouts of a two-bit adder were created for five different timing specifications. The adder circuit was composed of complex logic gates contained 52 unique transistors. The object was to create the smallest layout possible while maintaining an aspect ratio within 10% of 1.5 and meeting the timing specification.

Table 6.1: Area vs. speed measurements.

Delay (ns)	Freq. (MHz)	Active Area $W_t(\lambda)$	Total Area(λ^2)	Aspect Ratio	Total/ Active	Wr (λ)
15.98	62.6	995	50048	1.5	50.3	48
11.9	84.0	1133	57600	1.4	50.8	56
7.93	126.1	1373	72576	1.6	47.0	64
6.8	147.1	4329	154752	1.6	35.7	112
5.95	168.1	8163	228480	1.4	28.0	160

Figure 6.17 shows the total area versus the speed of the circuit created. As can be expected, the area of the circuit increases as the timing requirements of the circuit become more stringent. But even though the total area and W_t under the polysilicon lines increase as the frequency increases, the ratio of the total area over W_t does not increase. This ratio, although unconventional, provides a performance measure of the layout system. Table 6.1 presents all of the data collected.

iCGEN : A CMOS LAYOUT SYNTHESIS SYSTEM 241

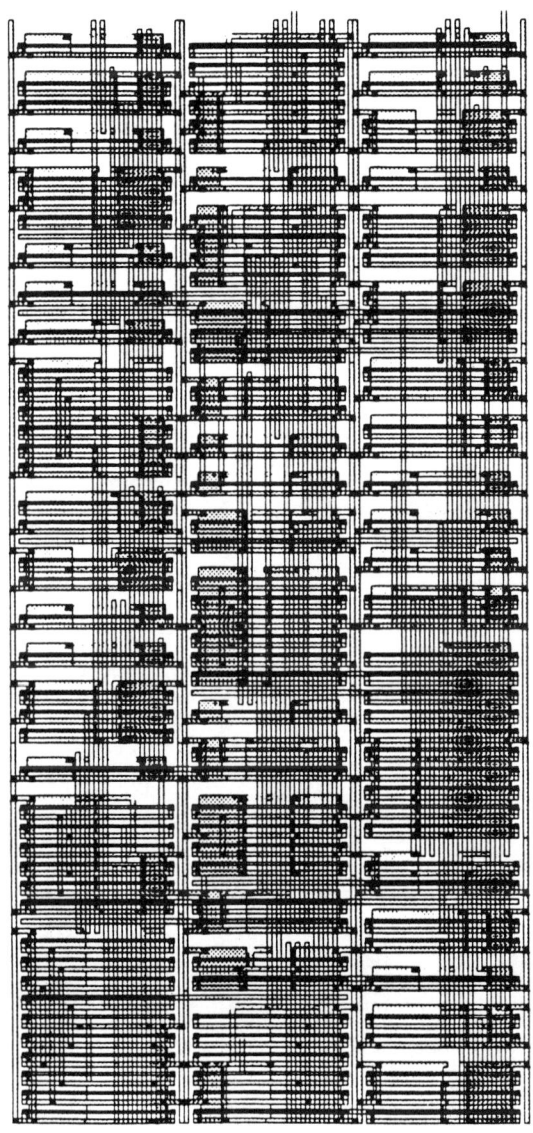

Figure 6.15: Automatic layout of industrial circuit 1.

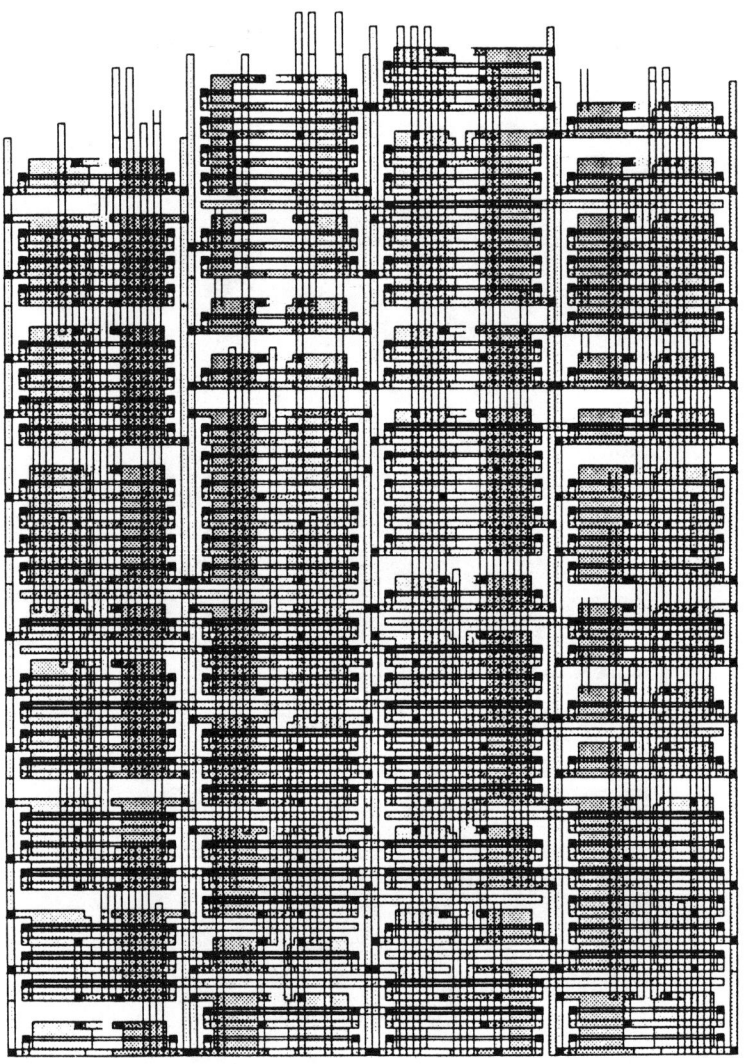

Figure 6.16: Automatic layout of industrial circuit 2.

iCGEN : A CMOS LAYOUT SYNTHESIS SYSTEM

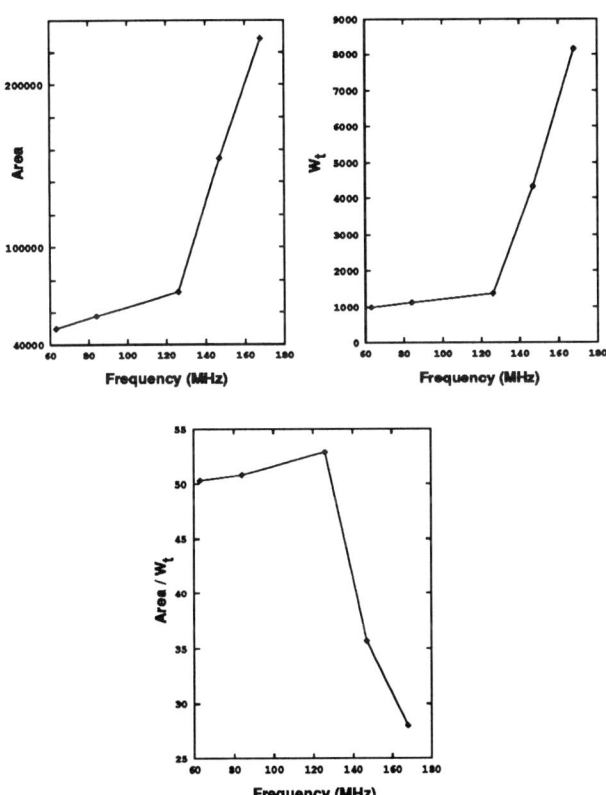

Figure 6.17: Area vs. speed for a complex two-bit adder circuit.

It can be noted that in this exercise, the transistor sizing is area-effective over the clock frequency range between 60 MHz to about 120 MHz, and tends to become expensive after reaching the knee at 120 MHz. Another important observation is that the ratio of the total chip area to the active area remains flat, which indicates the efficiency of the routing algorithms used.

6.4.10 Summary

In this section, we have presented a new high-performance CMOS layout synthesis tool. The layout of the design is row based and uses over the cell routing of intercell signals, i.e., no routing channels in between the rows. Based on the regular layout platform shown in Figure 6.13, the optimal row height and the number of rows can be calculated to minimize the total area of the design while meeting a specified aspect ratio. The modular nature of the system allows standard placement programs to be used to determine the relative placement of logic cells. Then, 0-1 integer linear programming techniques are employed to determine the optimal global routing of intercell signals. The actual layout generation of each cell utilizes graph theoretic algorithms to determine the optimal order of transistor gates to reduce the width of each cell. After the layout of the entire circuit is generated, the circuit and RC parasitics are extracted for the transistor sizing program, which uses convex optimization techniques to determine the optimal transistor sizes while meeting timing specifications. The transistor sizes are fed back to tune the circuit layout. This process can be integrated, without the user's intervention, until a stable layout is achieved that meets the timing specifications. The system has produced layouts which were smaller than industrial circuits manually created, even without geometrical compaction. The layout area was further reduced by 20% when the transistor sizes were optimized by using our convex programming approach.

Bibliography

[Bey78] W. H. Beyer, editor, *CRC Handbook of Mathematical Sciences*, CRC Press, 1978.

[BJ90] M. R. C. M. Berkelaar and J. A. G. Jess, "Gate sizing in MOS digital circuits with linear programming," *Proceedings of the 1990 European Design Automation Conference*, 1990.

[Boy88] D. G. Boyer, "Symbolic layout compaction review," *Proceedings of the 25th ACM/IEEE Design Automation Conference*, pages 383–389, June 1988.

[BP83] M. Burstein and R. Pelavin, "Hierarchical wire routing," *IEEE Transactions on Computer-aided Design of Integrated Circuits and Systems*, CAD-2:223–234, October 1983.

[CC89] M. Chang and C. Chen, "PROMPT3 : A cell-based transistor sizing program using heuristic and simulated annealing algorithms," *Proceedings of the 1989 Custom Integrated Circuits Conference*, pages 17.2.1–17.2.3, May 1989.

[Cha86a] P. K. Chan, "An extension of Elmore's delay," *IEEE Transactions on Circuits and Systems*, CAS-33(11):1147–1149, November 1986.

[Cha86b] P. K. Chan, "An extension of Elmore's delay and its application for timing analysis of MOS pass transistor networks," *IEEE Transactions on Circuits and Systems*, CAS-33(11):1149–1152, November 1986.

[Che88] H. Y. Chen, *Design Automation for High Performance CMOS VLSI Circuits*, Ph. D. thesis, University of Illinois at Urbana-Champaign, December 1988.

[Cir87] M. A. Cirit, "Transistor sizing in CMOS circuits," *Proceedings of the 24th ACM/IEEE Design Automation Conference*, pages 121–124, June 1987.

[CK89] P. K. Chan and K. Karplus, "Computing signal delay in general RC networks by tree/link partitioning," *Proceedings of the 26th ACM/IEEE Design Automation Conference*, pages 485–490, July 1989.

[CK91]   H. Y. Chen and S. M. Kang, "iCOACH: A circuit optimization aid for CMOS high-performance circuits," *Integration, the VLSI Journal*, 10(2):185–212, January 1991.

[CL75]   L. O. Chua and P.-M. Lin, *Computer-aided Analysis of Electronic Circuits: Algorithms and Computational Techniques*, Prentice-Hall, 1975.

[Cla83] F. H. Clarke, *Optimization and Nonsmooth Analysis*, Wiley-Interscience, 1983.

[CRS90] J. P. Cohoon, D. S. Richards, and J. S. Salowe, "An optimal Steiner tree algorithm for a net whose terminals lie on the perimeter of a rectangle," *IEEE Transactions on Computer-aided Design of Integrated Circuits and Systems*, 9(4):398–407, April 1990.

[CS89]   P. K. Chan and M. D. F. Schlag, "Bounds on signal delay in RC mesh networks," *IEEE Transactions on Computer-aided Design of Integrated Circuits and Systems*, 8(6):581–589, June 1989.

[Dai87] W.-M. Dai *et al.*, "BEAR: A new building-block layout system," *Proceedings of the 1987 International Confer-*

ence on Computer-aided Design, pages 34–37, November 1987.

[DA89] Z. Dai and K. Asada, "MOSIZ : A two-step transistor sizing algorithm based on optimal timing assignment method for multi-stage complex gates," *Proceedings of the 1989 Custom Integrated Circuits Conference*, pages 17.3.1–17.3.4, May 1989.

[Dag87] M. Dagenais, "Timing analysis for MOSFET's: An integrated approach," Technical Report TR-88-2R, Electrical Engineering, McGill University, June 1987.

[DD89] P. Duchene and M. J. Declercq, "A highly flexible sea-of-gates structure for digital and analog applications," *IEEE Journal of Solid-State Circuits*, 24(3):576–584, June 1989.

[DDK91] P. Duchene, M. J. Declercq, and S. M. Kang, "An integrated layout system for sea-of-gates module generation," *Proceedings of the 1991 European Design Automation Conference*, pages 237–241, February 1991.

[DFH89] A. E. Dunlop, J. P. Fishburn, D. D. Hill, and D. D. Shugard, "Experiments using automatic physical design techniques for optimizing circuit performance,"

Proceedings of the 32nd Midwest Symposium on Circuits and Systems, Urbana, IL, August 1989.

[DGR87] E. Detjens, G. Gannot, R. Rudell, A. Sangiovanni-Vincentelli, and A. Wang, "Technology mapping in MIS," *Proceedings of the 1987 International Conference on Computer-aided Design*, pages 116–119, November 1987.

[DGR92] M. R. Dagenais, S. Gaiotti, and N. C. Rumin, "Transistor-level estimation of worst-case delays in MOS VLSI circuits," *IEEE Transactions on Computer-aided Design of Integrated Circuits and Systems*, 11(3):384–395, March 1992.

[Eck80] J. G. Ecker, "Geometric programming : Methods, computations and applications," *SIAM Review*, 22(3):338–362, July 1980.

[Elm48] W. C. Elmore, "The transient response of damped linear networks with particular regard to wideband amplifiers," *Journal of Applied Physics*, 19, January 1948.

[EST90] Y. Enomoto, T. Sasaki, S. Tsutsumi, and S. Tone, "A 200K gate 0.8um mixed CMOS/BiCMOS sea-of-gates," *Proceedings of the IEEE International Solid-State Circuits Conference*, pages 92–93, February 1990.

[Eve79] S. Even, *Graph Algorithms*, Computer Science Press, 1979.

[FD85] J. P. Fishburn and A. E. Dunlop, "TILOS : A posynomial programming approach to transistor sizing," *Proceedings of the 1985 International Conference on Computer-aided Design*, pages 326–328, November 1985.

[Fis92] J.P. Fishburn, Private communication, 1992.

[Gal90] J. Gallia *et al*, "High-performance BiCMOS 100K-gate array," *IEEE Journal of Solid-State Circuits*, 25(1):142–149, February 1990.

[GH90] J. A. Gasbarro and M. A. Horowitz, "A single-chip, functional tester for VLSI circuits," *Proceedings of the IEEE International Solid-State Circuits Conference*, pages 84–85, February 1990.

[GL89] G. H. Golub and F. H. Van Loan, *Matrix Computations*, The Johns Hopkins University Press, 1989.

[GVL91] T. Gao, P. M. Vaidya and C. L. Liu, "A new performance-driven placement algorithm," *Proceedings of the 1991 International Conference on Computer-aided Design*, pages 44–47, November 1991.

[GZ92] D. S. Gao and D. Zhou, "Propagation delay in RLC interconnection trees," Submitted to *IEEE Transactions on Computer-aided Design of Integrated Circuits and Systems*, June 1992.

[Han66] M. Hanan, "On Steiner's problem with rectlinear distance," *SIAM Journal of Applied Mathematics*, 14:255–265, March 1966.

[Hed87] K. S. Hedlund, "AESOP : A tool for automated transistor sizing," *Proceedings of the 24th ACM/IEEE Design Automation Conference*, pages 114–120, June 1987.

[HF91] L. S. Heusler and W. Fichtner, "Transistor sizing for large combinational digital CMOS circuits," *Integration, the VLSI Journal*, 10(2):155–168, January 1991.

[HJ87] N. Hedenstierna and K. O. Jeppson, "CMOS circuit speed and buffer optimization," *IEEE Transactions* on *Computer-aided Design of Integrated Circuits and Systems*, CAD-6:270–281, March 1987.

[HK85] T. C. Hu and E. S. Kuh, "Theory and concepts of circuit layout," *VLSI Circuit Layout: Theory and Design*, IEEE Press, 1985.

[HKE89] B. Hoppe, O. Kiehl, V. Eisele, T. Huber, D. Schmitt-Landsiedel, and G. Neuendorf, "Posynomial delay models for optimisation-based transistor sizing in digital CMOS VLSI circuits," *Proceedings of the 1989 European Conference on Circuit Theory and Design*, pages 275–279, 1989.

[HNS90] B. Hoppe, G. Neuendorf, D. Schmitt-Landsiedel, and W. Specks, "Optimization of high-speed CMOS logic circuits with analytical models for signal delay, chip area, and dynamic power dissipation," *IEEE Transactions on Computer-aided Design of Integrated Circuits and Systems*, 9(3):236–247, March 1990.

[HS85a] T. C. Hu and M. T. Shing, "A decomposition algorithm for circuit routing," *VLSI Circuit Layout: Theory and Design*, IEEE Press, 1985.

[HS85b] T.C. Hu and M.T. Shing, "A decomposition algorithm for circuit routing," T.C. Hu and E.S. Kuh, editors, *VLSI Circuit Layout: Theory and Design*, pages 144–152, IEEE Press, 1985.

[HSF89] D. Hill, D. Shugard, J. Fishburn, and K. Keutzer, *Algorithms and Techniques for VLSI Layout Synthesis*, Kluwer Academic Publishers, 1989.

[HUH90] T. Hanibuchi, M. Ueda, K. Higashitani, M. Hatanaka, and K. Mashiko, "A bipolar-PMOS merged basic cell for 0.8um BiCMOS Sea-of-Gates," *Proceedings of the 1990 Custom Integrated Circuits Conference*, pages 4.2.1–4.2.4, May 1990.

[JK89] M. A. B. Jackson and E. S. Kuh, "Performance-driven placement of cell based IC's," *Proceedings of the 26th ACM/IEEE Design Automation Conference*, pages 370–375, June 1989.

[Kah87] M. Kahrs, "Matching a parts library in a silicon compiler," *Proceedings of the 1987 IEEE International Conference on Computer-aided Design*, pages 169–172, November 1987.

[Kan81] S. M. Kang, "A design of CMOS polycells for LSI circuits," *IEEE Transactions on Circuits and Systems*, 28(8):838–843, August 1981.

[Kan87] S. M. Kang, "Metal-metal matrix (M^3) for high-speed MOS VLSI layout," *IEEE Transactions on Computer-aided Design of Integrated Circuits and Systems*, 6(5):886–891, September 1987.

[Kan90] S. M. Kang, "Performance-driven layout of CMOS VLSI circuits," *Proceedings of the 1990 International Symposium on Circuits and Systems*, pages 881–884, May 1990.

[KB80] B. Kolman and R. E. Beck, *Elementary Linear Programming with Applications*, Academic Press, 1980.

[KC66] T. I. Kirkpatrick and N. R. Clark, "PERT as an aid to logic design," *IBM Journal of Research and Development*, volume 10, pages 135–141, March 1966.

[KCh90] S. M. Kang and H. Y. Chen, "A global delay model for domino CMOS circuits," *International Journal on Circuit Theory and Applications*, 18(3):289–306, May 1990.

[KGV83] S. Kirkpatrick, G. D. Gelatt, Jr., and M. P. Vecchi, "Optimization by simulated annealing," *Science*, 220(4598):671–680, May 1983.

[KKL87] K. Keutzer, K. Kolwicz, and M. Lega, "Impact of library size on the quality of automated synthesis," *Proceedings of the 1987 IEEE International Conference on Computer-aided Design*, pages 120–123, November 1987.

[KOI92] S. Kim, R. M. Owens, and M. J. Irwin, "High performance CMOS module synthesis with PERFLEX," *Proceedings of the International Workshop on Layout Synthesis*, pages 245–253, May 1992.

[Lau87] U. Ph. Lauther, "Top down hierarchical global routing for channelless gate arrays based on linear assignment," *VLSI '87*, pages 141–151, 1987.

[Lek90] N. Lek, "A global routing scheme for Sea-of-Gates arrays using zero-one optimization methods," Master's thesis, University of Illinois at Urbana-Champaign, 1990.

[LKT87] F. Lai, S. M. Kang, and T.N. Trick, "A timing error corrector for VLSI synchronous path circuits," *Pro-

ceedings of the 1987 IEEE International Conference on Computer-aided Design, pages 48–51, 1987.

[LL80] A. D. Lopez and H. F. S. Law, "A dense gate-matrix layout method for MOS VLSI," *IEEE Transactions of Electron Devices*, 27(8):1671–1675, August 1980.

[LM84] T.-M. Lin and C. A. Mead, "Signal delay in general RC networks," *IEEE Transactions on Computer-aided Design of Integrated Circuits and Systems*, 3(4), October 1984.

[Lue84] D. G. Luenberger, *Linear and Nonlinear Programming*, Addison-Wesley, 1984.

[Man65] G. Mangasarian, "Pseudo-convex functions," *SIAM Journal of Control and Optimization*, 3(2):281–290, 1965.

[Mar86] D. Marple, "Performance optimization of digital VLSI circuits," Technical Report CSL-TR-86-308, Stanford University, October 1986.

[Mar89] D. Marple, "Transistor size optimization in the Tailor layout system," *Proceedings of the 26th ACM/IEEE Design Automation Conference*, pages 43–48, June 1989.

[Mat85] M. Matson, *Macromodeling and Optimization of Digital MOS VLSI Circuits*, Ph. D. thesis, MIT, February 1985.

[MC80] C. Mead and L. Conway, *Introduction to VLSI Systems*, Addison-Wesley, 1980.

[MG86] D. Marple and A. El Gamal, "Area-delay optimization of programmable logic arrays," *Fourth MIT Conference on VLSI*, pages 171–194, April 1986.

[MG87] D. Marple and A. El Gamal, "Optimal selection of transistor sizes in digital VLSI circuits," *Stanford Conference on VLSI*, pages 151–172, 1987.

[MS84] M. Marek-Sadowska, "Global router for gate array," *Proceedings of the 21st ACM/IEEE Design Automation Conference*, pages 332–337, June 1984.

[MSH88] D. Marple, M. Smulders, and H. Hegen, "An efficient compactor for 45° layout," *Proceedings of the 25th ACM/IEEE Design Automation Conference*, pages 396–402, June 1988.

[Mur64] B. T. Murphy, "Cost-size optima of monolithic integrated circuits," *Proceedings of the IEEE*, 52:1537–1545, December 1964.

[Nag75] L. W. Nagel, "SPICE2: A computer program to simulate semiconductor circuits," Technical Report Electronics Research Laboratory Report # ERL-M520, University of California, Berkeley, May 1975.

[Noy77] R. N. Noyce, "Microelectronics," *Scientific American*, 237:62–69, September 1977.

[NRT87] A. P.C. Ng, P. Raghavan, and C. D. Thompson, "Experimental results for a linear program global router," *Computers and Artificial Intelligence*, pages 229–242, 1987.

[Oka89] M. Okabe *et al.*, "A CMOS Sea-of-Gates array with continuous track allocations," *Proceedings of the IEEE International Solid-State Circuits Conference*, pages 180–181, February 1989.

[Ous85] J. K. Ousterhout, "A switch-level timing verifier for digital MOS VLSI," *IEEE Transactions on Computer-aided Design of Integrated Circuits and Systems*, 4(3):336–349, July 1985.

[PR90] L. T. Pillage and R. A. Rohrer, "Asymptotic waveform evaluation for timing analysis," *IEEE Transactions* on *Computer-aided Design of Integrated Circuits and Systems*, 9(4):352–366, April 1990.

[PT89] T.-M. Parng and R.-S. Tsay, "A new approach to sea-of-gates global routing," *Proceedings of the 1989 International Conference on Computer-aided Design*, pages 52–55, November 1989.

[PTM71] E. Polak, R. Trahan, and D. Q. Mayne, "Combined phase I-phase II methods of feasible directions," *Mathematical Programming*, 17(1):61–73, 1971.

[Rao85] V. B. Rao, *Switch-level Timing Simulation of MOS VLSI Circuits*, Ph. D. thesis, University of Illinois at Urbana-Champaign, January 1985.

[ROT89] V. B. Rao, D. V. Overhauser, T. N. Trick, and I. N. Hajj, *Switch-Level Timing Simulation of MOS VLSI Circuits*, Kluwer Academic Publishers, 1989.

[RPG77] A. E. Ruehli, P. K. Wolff, Sr., and G. Goertzel, "Analytical power timing optimization techniques for digital systems," *Proceedings of the 14th ACM/IEEE Design Automation Conference*, pages 142–146, 1977.

[RPH83] J. Rubenstein, P. Penfield, and M. A. Horowitz, "Signal delay in RC tree networks," *IEEE Transactions* on *Computer-aided Design of Integrated Circuits and Systems*, CAD-2(3):202–211, July 1983.

[RTH83] V. B. Rao, T. N. Trick, and I. N. Hajj, "A table-driven delay-operator approach to timing simulation of MOS VLSI circuits," *Proceedings of the 1983 International Conference on Computer Design*, pages 445–448, 1983.

[Sak83] T. Sakurai, "Approximation of wiring delay in MOS-FET LSI," *IEEE Journal of Solid-State Circuits*, 18(4), August 1983.

[SCK91] A. Srinivasan, K. Chaudhary and E. S. Kuh, "RITUAL: A performance-driven placement algorithm for small cell ICs," *Proceedings of the 1991 International Conference on Computer-aided Design*, pages 48–51, November 1991.

[SFD88] J. Shyu, J. P. Fishburn, A. E. Dunlop, and A. L. Sangiovanni-Vincentelli, "Optimization-based transistor sizing," *IEEE Journal of Solid-State Circuits*, pages 400–409, April 1988.

[Shy88] J. Shyu, "Performance optimization of integrated circuits," Technical Report UCB/ERL M88/74, UCB, November 1988.

[SMN88]   Y. Suehiro, D. Miura, M. Naitoh, S. Tsutsumi, and T. Shirato, "A 120K-gate usable CMOS sea-of-gates packing 1.3M transistors," *Proceedings of the 1988 Custom Integrated Circuits Conference*, pages 20.5.1–20.5.4, May 1988.

[Sou79]   J. Soukup, "Global router," *Proceedings of the 16th ACM/IEEE Design Automation Conference*, pages 206–209, 1979.

[SR90]   S. S. Sapatnekar and V. B. Rao, "iDEAS : A delay estimator and transistor sizing tool for CMOS circuits," *Proceedings of the 1990 Custom Integrated Circuits Conference*, pages 9.3.1–9.3.4, May 1990.

[SSV85]   C. Sechen and A. Sangiovanni-Vincentelli, "The timberwolf placement and routing package," *IEEE Journal of Solid-State Circuits*, April 1985.

[SVR91]   S. S. Sapatnekar, P. M. Vaidya, and V. B. Rao, "A convex programming approach to transistor sizing for CMOS circuits," *Proceedings of the 1991 IEEE International Conference on Computer-aided Design*, pages 482–485, 1991.

[Tak89] T. Takahashi et al., "A 1.4M-transistor CMOS gate array with 4ns RAM," *Proceedings of the IEEE International Solid-State Circuits Conference*, pages 178–179, February 1989.

[Tha92] R. W. Thaik, "iCGEN: A compact timing-driven CMOS integrated circuit layout generator," M. S. thesis, University of Illinois at Urbana-Champaign, April 1992.

[Tim87] Yale University, *User's Guide to the TimberWolfSC Standard Cell Placement and Global Routing Package*, October 1987.

[TS77] H. Taub and D. Schilling, *Digital Integrated Electronics*, McGraw-Hill, 1977.

[Tuc84] A. Tucker, *Applied Combinatorics*, John Wiley & Sons, 1984.

[UV81] T. Uehara and W. M. VanCleemput, "Optimal layout of CMOS functional arrays," *IEEE Transactions on Computers*, 30(5):305–312, May 1981.

[Vai89] P. M. Vaidya, "A new algorithm for minimizing convex functions over convex sets," *Proc. IEEE Foundations of Computer Science*, pages 332–337, October 1989.

[Vai90]   P. M. Vaidya, "An algorithm for linear programming which requires $O(((m+n)n^2+(m+n)^{1.5}n)l)$ arithmetic operations," *Mathematical Programming*, 47:175–201, 1990.

[VEH90]   H. Veendrick, D. van den Elshout, D. Harberts, and T. Brand, "An efficient and flexible architecture for high-density gate arrays," *Proceedings of the IEEE International Solid-State Circuits Conference*, pages 86–87, February 1990.

[Won88]   A. Wong *et al.*, "A high density BiCMOS direct drive array," *Proceedings of the 1988 Custom Integrated Circuits Conference*, pages 20.6.1–20.6.3, 1988.

[Wya85]   J. L. Wyatt, "Signal delay in RC mesh networks," *IEEE Transactions on Circuits and Systems*, 32(5):507–510, May 1985.

[Wya87]   J. L. Wyatt, Jr., "Signal propagation delay in RC models for interconnect," A. E. Ruehli, editor, *Circuit Analysis, Simulation and Design*, pages 254–291, North-Holland, 1987.

[Yos89]   H. Yoshimura *et al.*, "500K transistor custom BiCMOS LSI using automated macrocell design," *Proceedings of the IEEE International Solid-State Circuits Conference*, pages 122–123, 1989.

[ZOO87] XMP Software, Inc., *User's Manual For ZOOM/XMP*, July 1987.

Index

A

AESOP 101-102, 111

B

Blocking transistor 90, 92, 132

C

CATS 106-108

Channel-connected components
  see Components

Cirit's method 101, 110

Combinational subnetwork extraction 52-56

Components 31-33

Convex programming 62, 114-133, 204

Critical transistor 89, 91-92, 132

E

Elmore delay 21-28, 67-73, 87-88

Euler paths 235

G

Gate array 5-6

Generalized gradient 96-100

H

Hold time 54-56

I

IC design process 2-4

iCGEN 192-193, 216-245
  intercell signal routing 231-234
  layout platform 222-226
  logic cell placement 230-231
  logic cell layout 234-236
  optimal row height 227-230

results 238-244

transistor size optimization 236-238

iCOACH 106-108

iCONTRAST 62-77, 113-139, 221, 236-238

   convex optimization algorithm 115-133

   in iCGEN 221, 236-238

   results 133-140

   timing analyzer 62-77

iDEAS 109

ILP see Zero-one integer linear programming

Inverter delay modeling 41-48, 64-65, 197-202

L

Largest resistive path 68-73

Lagrangian multiplier approaches 101-106, 111-112

LRP see Largest resistive path

M

M^3 192, 213-216

Macromodeling 14-15, 31-50

   analytical method 40-50

   table look-up 33-40

Marple's approach 102-106, 111, 138

Method of feasible directions 93-100, 110

Micromodeling 14, 16-30, 63-65

MOGLO 108

MOSIZ 106-108

N

Nonstep inputs 74-76

P

Penfield-Rubenstein bounds 28-30

PERT 56-60, 66

PLA 4-5

Polytope center computation 122-131

INDEX

Posynomial 62, 78, 88, 113-114, 204

Preconditioned conjugate gradient method 126-127, 129-130

PROMPT3 108

R

RC mesh 18

RC models

    for transistors 19, 63-65

    for interconnect 19-21

RC tree 16-18

S

Sea-of-gates arrays 6-7, 137-138, 218-219

Set-up time 54-55

Shyu's approach *see* Method of feasible directions

SOG global router

    phase one routing 145, 146-150

    phase three routing 146, 156-169

    phase two routing 145, 150-155

    results 169-183

Standard cells 7-8

Standard cell design (CMOS) 192, 193-212

Supporting transistor 90, 92, 132

T

TILOS 85-93, 110

Timing-driven layout 9-10

Transistor sizing problem definition 84-85

W

Worst-case delay estimation 50-52, 62-79

Z

Zero-one integer linear programming 141-189, 220, 231-234